STREETSCAPES

HOW TO DESIGN AND DELIVER GREAT STREETS

COLIN J DAVIS

© Colin J Davis, 2018

Published by CJDA Ltd.
Registered in England number 03145434

ISBN 978-0-9931004-1-3
The right of Colin J Davis to be identified as the Author of
this Work has been asserted in accordance with the
Copyright, Designs and Patents Act 198 sections 77 and 78.

British Library Cataloguing—in—Publication Data
A catalogue record for this book is available from the British
Library.

Graphics and design: William Spencer and Colin Davis

While every effort has been made to check the accuracy and
quality of the information given in this publication, neither
the Author nor the Publisher accept any responsibility for
the subsequent use of this information, for errors or
omissions that it may contain, or for any misunderstandings
arising from it.

www.publicrealm.org
www.streetscapes.online

FOREWORD

I never imagined as a first year architectural student that one page of one issue of the *Architectural Review* would influence much of what I did in my subsequent career. The sketches of streetscapes by Gordon Cullen I saw then still make sense and so they are the introduction to this book.

Streetscapes are central to all our lives. Putting Cullen's ideas into practice is what very many people spend enormous efforts trying to achieve. But although they are seen and understood as a whole, streetscapes are delivered in many, sometimes uncoordinated, parts. The key to success seems to be to bring all those involved more closely together.

Much of the book is a distillation of ideas and experience shared with colleagues in many disciplines. They are thanked on page 140. A pictorial, straightforward format has been chosen to help people in one discipline understand more fully the objectives and challenges of others.

The aim quite simply is to encourage greater collaboration so that we can all help create even better streetscapes.

Colin J Davis RIBA MRTPI FCIHT March 2018
For all enquiries: PublicRealmPRIAN@aol.com

CONTENTS

INTRODUCTION

"Streets are places too" is a phrase that sums up this book. We all use streets everyday for getting about, but also as places where we live and work. There are streets we like and those we don't; streets that are attractive, welcoming and convenient, and there are those that are not. From this follows the question: could those that are not, be made better?

To help readers new to the subject we use where possible everyday terms defined by standard dictionaries. Streetscape is the view or scene of streets. A street is a public road in a city, town or village, typically with houses and buildings on one or both sides, and a road is the part of a street that is used by vehicles.

A first step is to be clear as to the problems that need to be remedied, to apply objective assessments and consider what can be done. It is easy to find press stories that point out opinions that our streets are untidy, inefficient, dangerous, inconvenient and expensive.

The book addresses the problems of improving street design by exploring five theoretical principles or objectives of good street design and how they can be – and have been – applied in practice. To select five groups of objectives emphasises the complexity of street design. Street design decisions are seldom binary: either/or. A good design would meet all five design objectives. And as many of the individual design objectives tend to be the province of a discrete group of specialists or separate professional disciplines, good design involves the co-ordination of all the disciplines.

EFFECTIVE STREETSCAPES: GORDON CULLEN

The idea that more than one discipline is involved in the creation of an enjoyable yet practical public realm is seen in the work of the streetscape expert Gordon Cullen. In December 1956 the *Architectural Review* published a special issue titled Counter Attack. It was its own response to a previous issue of the magazine that had been devoted to the parlous state of the built environment.

One article was little more than a series of simplified pen-and-ink sketches of both town and country street scenes, accompanied only by very brief notes. It was drawn and written by Gordon Cullen, who had been responsible for illustrating post-war improvement schemes and was interested in how streetscape could be appreciated by both a static and moving observer. The effect on someone as they moved from one space to another in a sequence of experiences fascinated him.

He was also very conscious of the clear differences between city, town, suburb and countryside, which had been pointed out by John Burns at the beginning of the twentieth century.

Cullen's sketches in the *Architectural Review* drew on this knowledge, and explained by means of before and after views of how a typical urban street and country lane could be visually improved. The notes under the urban scene suggested how the buildings could be renovated. They showed a city street suffering from bomb damage still remaining after the Second World War (Fig. 0.01). The top of a church spire is missing, and there are gaps where buildings have been razed to the ground. Large, ugly advertisement signs have been fixed to prominent historic buildings, and the road itself has an accumulation of sign clutter and inappropriate municipal planting on an over-elaborate traffic roundabout with hideous street furniture.

In the sketch showing the same scene after improvement (Fig. 0.02) the empty gaps have been filled, historic buildings have been repaired and the ugly signs have been removed. In the intervening years, improvements such as these have become common practice as they are carried out or encouraged through current planning and conservation legislation. But the sketches also show the roundabout replacing with a simplified road system to create a new public space.

0.01 Typical urban street of the 1950s: semi-derelict buildings, sign clutter and over-elaborate traffic arrangements

0.02 The same street: repaired buildings, revised traffic systems and a new public space

Cullen applies the same principles to the sketches of a country scene. The advertisement signs, ugly overhead wires and rigid pavement lines (Fig. 0.03) are removed and replaced by landscape in the background, and a softer more rural treatment of the road (Fig. 0.04). The most apparent result is that there is a greater distinction between the urban and rural.

Today, some sixty years after the sketches were published, the scenes in our streets are obviously different but the points made are still relevant. Though most towns do have traffic-free zones, many ordinary streets have a visual disorder caused by little co-ordination of the standard street furniture or traffic equipment. Cullen's captions explained what he considered should be done; he appears to have assumed that there was a co-ordinating body or organisation that would be able to carry out his suggestions. Unfortunately that has not happened, primarily because town planning legislation is not concerned with the day-to-day decisions regarding the design and maintenance of all that goes on in the public highway. However, though there are no formal co-ordination procedures, the fact that in some towns and cities co-ordination has been achieved indicates that with some application and effort improvements can be made.

0.03 Typical 1950s rural scene: overhead wires, hoardings and visually rigid roads

0.04 The same rural scene: clutter and rigidity removed, replaced by landscape and visually sensitive road details

Part I
FIVE DESIGN OBJECTIVES

THE FIVE DESIGN OBJECTIVES OF A SUCCESSFUL STREET

Gordon Cullen's forward-thinking sketches brilliantly combine concerns about the quality of a street as a place, with recognition of the practicalities of efficient movement. Since the sketches were published in the mid 1950's there are three more groups of design objectives that must also be considered. With eight times the number of cars on the road and the natural horror of an increase in danger, the science of accident reduction and road safety has become far more important. Secondly, the diversity of activities that take place in streets has increased. Many need special equipment or adaptations of the design of the street in order to function, such as arrangements for public transport and cyclists, support for disabled people and systems for parking, as well as seamlessly incorporating landscape, markets and entertainment. Finally, funding arrangements and viability are paramount as unless they are in place nothing will happen. The positive or negative quality of a street has a direct influence on economic wellbeing and this is recognised by businesses.

The theme of this book therefore is that five key design objectives need to be (and *can* be) met in order to achieve a successful street:

1. Attractive places
2. Efficient movement
3. Road safety
4. User-friendly design
5. Funding and commercial viability.

The first part of this book examines each of these five objectives in turn, explaining what they each mean and how best practice can be extrapolated from them. The second part of the book presents six case studies that show how the design objectives outlined in Part 1 have been incorporated into completed schemes. Drawn from different parts of the country, they include urban, suburban and rural contexts.

A COLLABORATIVE EFFORT

All successful streets are the product of interdisciplinary effort and specialist expertise. Consequently, a key theme of this book is collaboration. A specialist may be a member of a design or maintenance team and may only be dealing with a single facet of the total process. To help put all this into context, the book brings together the constituent parts of the street design process.

"What do we mean by the 'public realm'? Well, it's the bit we all share. It's the open park, the sheltered street, the pedestrian square, the quiet enclave or tree-lined verge. These are the places that shape our feelings of wellbeing. This is our shared kingdom. Our collective realm."

Griff Rhys Jones
President, Civic Voice

"Our Review is a rallying call to heighten awareness of what and can be done. There is a need for a cross-cutting commitment to make the ordinary better and to improve the everyday built environment. Essentially this includes streetscapes and the closer attention to how our ordinary streets are designed and managed."

Sir Terry Farrell CBE
Chair, National Review of Architecture and the Built Environment

1.01 The countryside, inspiring and spacious. Surrey Hills Area of Outstanding Natural Beauty

ATTRACTIVE PLACES

"Consult the genius of the place," the poet Alexander Pope urged his friend the Earl of Burlington in 1731. His advice holds true today: landscape design, or the making of a place, should always be adapted to the context in which it is located. This has led to the use of the term "Place" as shorthand for the unique character of a particular location that should be respected and, if possible, enhanced in any physical works that are intended to make somewhere more attractive.

Pope explained how to identify the first of the five objectives that we consider essential for a successful street: an attractive place. He went on to list the features to consider: the hills, woods, rivers, the sense of scale, and the resulting human emotions and responses.

We accept that a place can be inspiring, spacious and exhilarating, somewhere to go for recreation (Fig. 1.01) but a place can be forbidding, dark and dingy: somewhere to avoid (Fig. 1.02). Or it can be welcoming, comfortable and reassuring, where we might like to live (Fig. 1.03).

In addition to serving a clear purpose and having a practical convenience, there are many characteristics that can make a place attractive: the pleasant contrasts between a city centre and a rural village, pleasant memories triggered by old buildings. Some places have been specially designed to be attractive with prestigious buildings or purposely designed outdoor spaces and streetscapes. In this chapter we first examine how streets, including roads and lanes, can contribute to and enhance the distinct streetscape attractiveness of a city, a town, a suburb or the countryside. We then consider how some principles can be applied to increase the attractiveness of places that have high volumes of traffic.

1.02 Somewhere to avoid

1.03 Somewhere we might like to live: Bedford Square, a Georgian square, London

1.04 The city dignified: Carlton House Terrace, London

Identifying the character of a place

Having accepted that a place has character, it is helpful to put some order into understanding what *is* the character of a place rather than a simple progressive hierarchy of good, better and best. Places are different: a city centre can be dynamic and, in contrast, the countryside can be relaxing. Both can be excellent, and to experience the contrast can be a great pleasure. The political thinker John Burns described in the early twentieth century what he felt were the appropriate attributes of different categories of place. The city, he said, should be dignified (Fig. 1.04), the town, pleasant (Fig. 1.05) and the suburb, salubrious (Fig. 1.06). He could have added that the countryside should be inspiring.

Each category has a value. One is not necessarily better than another, yet there are grades of quality within each category. Cullen pointed out what he saw as a tendency for everywhere to look the same as he felt the distinctions between city and countryside were being eroded. However, the wide range of materials and construction techniques now available allow the design of a street, road or lane to enhance the specific character of a place so that a city becomes more dignified, a town more pleasant, a suburb more salubrious and the countryside more inspiring. Our case studies assess examples of each category from the City of London to Bibury village.

1.05 The town pleasant: Aylesbury

1.06 The suburb salubrious: Hampstead Village

Recording the memorable qualities of a place in a streetscape analysis

Having recognised the category of a particular place's character as suggested by John Burns: city, town, suburb or countryside, the next step is to analyse the memorable streetscape qualities of the place within its category. If it is in a city centre we ask what makes it seem dignified. If it is in a town we ask what makes it pleasant, and so on with regard to salubrious suburbs or inspiring countryside. In simple terms it is what all visitors say to themselves when they walk down a street for the first time: "What do I like about the place?" Here to help make the streetscape analysis more considered we list some of the things to note, possibly as an annotated rough sketch: the historic and prominent buildings as a landmark or in groups, the spaces between buildings, including roads, and the views and vistas into and out of spaces (Fig.1.07). Every analysis sketch will be different, as it may focus on a small town market place, suburban road or village street.

HISTORIC BUILDINGS

Most people appreciate and even love historic buildings. They are a direct link to what has taken place at a particular location, and they help people to relate to their own past. As they are all around us and are easily accessible, they provide a setting for all the activities that take place now.

1.07 A streetscape analysis includes significant historic buildings, spaces between buildings, views and distant vistas

1.08 Traditional building with its prestige entrance for people directly from the street and a vehicle access at the side

1.09 Life of the building spills out on to the pavement

Historic buildings can form the sense of continuity that is a valuable ingredient in new developments. For example, a Victorian chapel or church hall may be converted into flats, as might be a public house, barn or old school. But though the purpose of the buildings may have changed, the physical form might still be easily recognised and appreciated, and are a reminder of how people once lived.

Churches are still dominant as structures in most town centres. Medieval churches, having been built as Roman Catholic churches, were adapted after the Reformation but still retain their prominent locations. Though not used by so many regular parishioners, they are still important to the community on many occasions, especially when local people feel the need to come together. Many Georgian inns have remained simply because they are in good commercial locations, their buildings have been adapted to a restaurant to meet changing tastes and the former stable yard may have been converted into a car park, but the same built form and external appearance will have been retained. Victorian town halls and libraries were well constructed and built with local pride; typically they contain a wealth of decoration and art that relates to local events and personalities, and very often they still have a practical use so that they remain largely unchanged externally, in good condition and as intended at prominently position in the streetscape. Noting where these buildings are in a street is a good way to start to understand their historical context.

RELATIONSHIP OF BUILDINGS TO ROADS AND STREETS

There is a long history of domestic buildings being designed so that people and goods can enter straight from the street. Typically this puts the focus of a building on the place where people enter, and so any decoration is around the entrance – for prestige – while carts are offloaded from a passageway to the rear (Fig. 1.08). Similarly the direct use of a building in conjunction with the public space immediately outside is quite common, and can be seen where a shop or cafe has sales or a sitting space on or next to the public pavement (Fig. 1.09). This interaction of building with public pavement is important to the streetscape. It can vary according to the size of the building or groups of buildings and the size and capacity of the road.

15

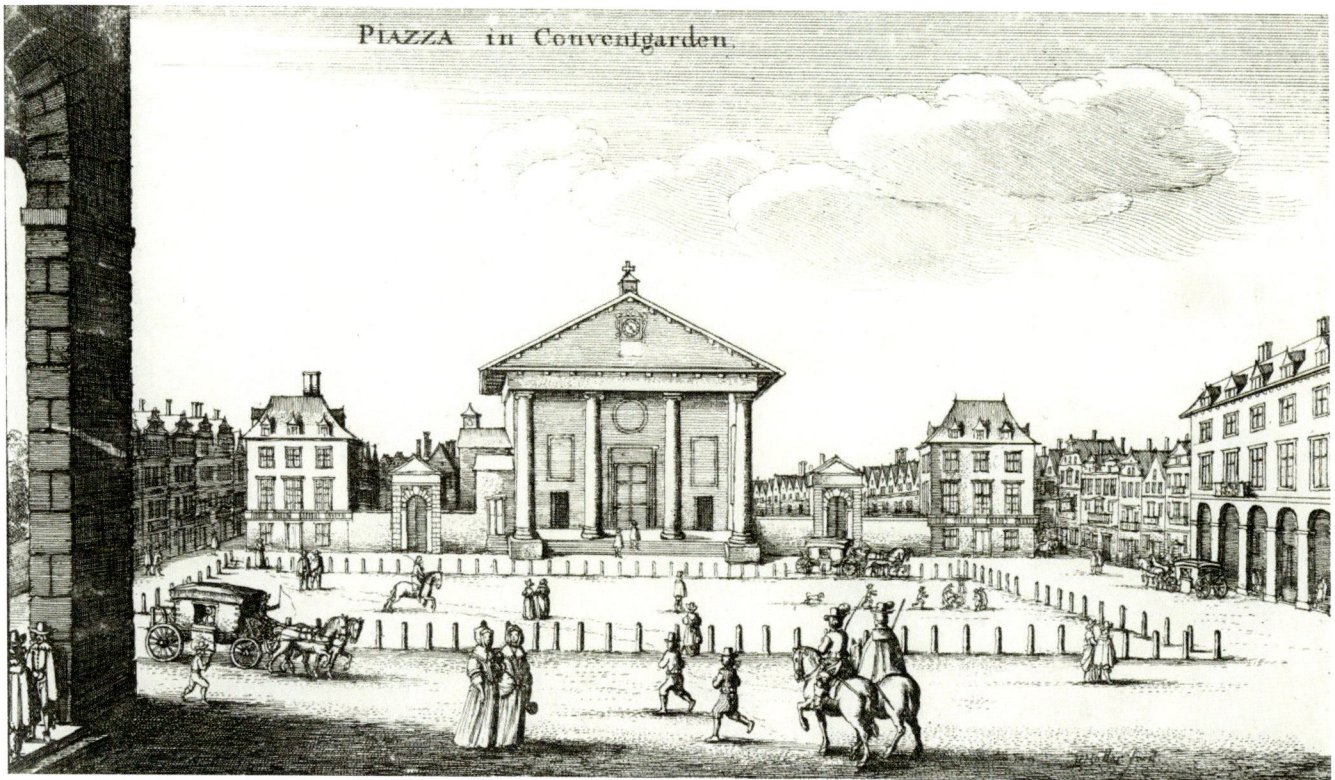

1.10 The Piazza, Covent Garden, built in the 1630s, was the first commercially financed public square in the UK

Buildings and roads: outdoor spaces

Buildings can also be arranged to form interesting or impressive outdoor spaces such as squares, crescents, circuses or avenues. When the powers of land ownership have been combined with design skills it has been possible to arrange groups of prestigious buildings, open spaces and roads in a comprehensive design of linked spaces. A commercially financed public square was first carried out in England by the Duke of Bedford to the designs of Inigo Jones in the 1630s at the Piazza, Covent Garden (Fig. 1.10), examined further in Chapter 5.

From then the concept was developed notably in Bath by John Wood the elder in 1725 at Queen Square (Fig. 1.11). The prestigious townhouses remain largely as they were originally built around a rectangle of roads and the gardens at Queen Square, linked by the terraces of Gay Street to a plantation of trees at The Circus, then through Brock Street to the Royal Crescent, built in 1767-75, with its spacious sweep and stunning views.

16

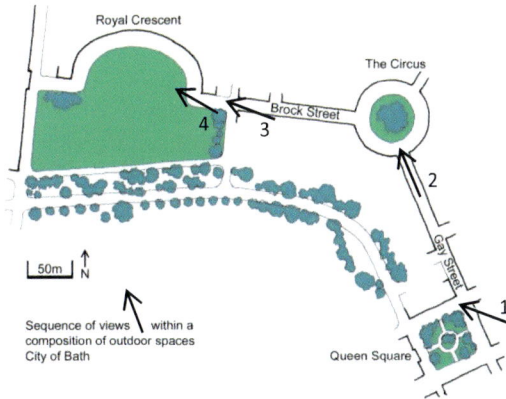

1.11 Sequence of views within a composition of outdoor spaces, City of Bath

1 Queen Square. Individual houses made to look like a grand palace

2 The Circus, inspired by the Colosseum, Rome, though facing inward rather than outward

3 The Royal Crescent is approached from the side...

4 so that its full grandeur is a surprise

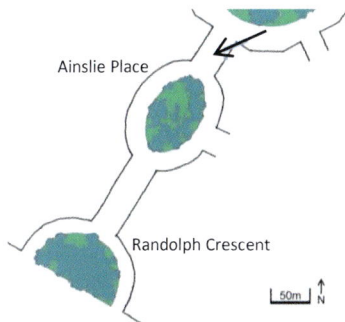

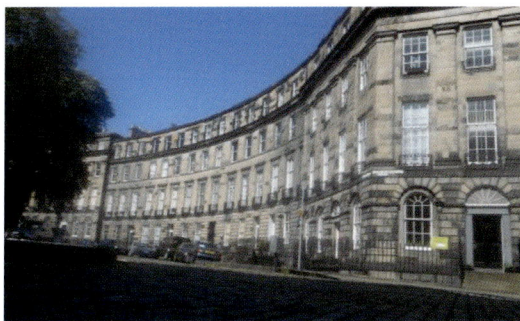

1.12 Composition of outdoor spaces, Edinburgh

The design of the facades of each group of houses as well as their visual relationship to the adjacent roads were strictly co-ordinated to create a single comprehensive composition incorporating a progression of carefully controlled views and vistas into and out of the public spaces. Here the high status of these groups of buildings and public spaces is emphasised by them being adjacent to more modest streets, courts, mews and alleys, designed and built at the same time for supporting roles at a smaller scale and with less expensive materials.

The New Town of Edinburgh was laid out by James Craig in the 1760s. The first part was a simple regular grid with lavish Charlotte Square and St Andrews Square at each end of the wide, central George Street. Open landscaped gardens on each side complete the layout. This was followed in the 1820s by a more innovative extension to the New Town (Fig. 1.12) at Randolph Crescent, Ainslie Place and Moray Place, which is a sequence of formal intricate spaces. A series of innovatively linked oval-shaped and circular gardens surrounded by roads are contained within curved facades of stone-faced classically designed buildings. The roads in each case are part of the composition.

A sequence of informal spaces & vistas

Having examined the UK's rich heritage of designed public spaces it is interesting to look at a configuration of streets and places that form a series or sequence of views, vistas and enclosed spaces that were not necessarily consciously or totally designed. Nevertheless they create a memorable experience and an excellent example is at Grainger Town, Newcastle, where it is possible to discover the subtle variations and contrasts as well as the ever-changing sequence of vistas that can be experienced in a short urban walk (Figs 1.13). It starts at the well-known Grey Street, turns off into a narrow street known as High Bridge, continues into Cloth Market and finishes in the stable yard of the Old George Inn. Walking and seeing the changes and contrasts is rewarding. Yet the distance is quite short, less than 150 metres.

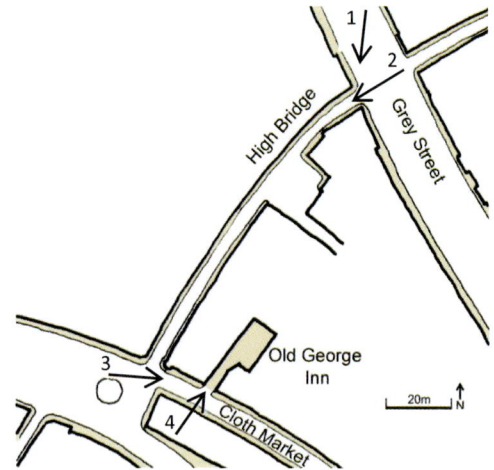

1.13 The sequence of spaces and vistas from Grey Street to the stable yard of the Old George Inn

The sequence is:

1 Grand prestigious shops and buildings of Grey Street

2 The intriguing High Bridge draws people on

3 More modest buildings of Cloth Market with inviting shops and doorways

4 Through an old cartway lies the intimate courtyard and the Inn. A place to rest and relax

1.14 The walk along the elevated walkway is the same distance as from Grey Street to the Old George Inn

1 A long boring walkway with little to look at apart from the backs of buildings and the traffic

2 At many places people walk on long detours

At each step there is some sort of interest – in terms of architecture, streetscape, a new vista and also in terms of the commercial and human activity. All the streets are both enjoyable routes to other places and destinations in themselves. The design of the road and pavement surfaces is an important element in the total experience of the walk. It can relate visually to each building in each street, and help to emphasise the distinct character of the individual streets and spaces.

But instead of helping people to appreciate a place, roads can become barriers. Since the mid-twentieth century, a system of high-speed roads and motorways has been built. There is a tendency for their design to ignore the quality of places they pass through, especially where they enter urban areas. It is argued that as no one lives, works or walks along these roads they are less important as places. However, their physical and visual impact is quite considerable, and the extent of land needed for a simple but high-speed road junction is as great as a whole city neighbourhood. They also create significant physical barriers for people trying to get about on foot.

CROSSING A MOTORWAY ON FOOT

A typical example is also at Newcastle where the A167(M) road enters the urban area of the city. As well as taking a large area of land, the road has the effect of dividing one part of the city from the other. The plan (Fig. 1.14), which is to the same scale as the Grey Street plan, shows that the distance across the road and motorway complex from Minden Street at a student accommodation shop westwards to the next building, which is at Market Street, is about 200 metres. Elevated walkways and bridges are provided, but are not at all welcoming or convenient for people on foot. The route is uncomfortably windy and exposed to the elements. It is not direct or easy for a pedestrian to understand, and many parts of the elevated walkway system require people to travel in a different direction to that intended. It is boring and, apart from the traffic, there is nothing of interest to look at, except possibly some distant landscape on a roundabout. The adjacent buildings, rather than facing the road and relating to it, present their unattractive facades towards the road, rather like buildings strung along a railway line.

Main roads in town and countryside

Fortunately not all motorways cut uncompromisingly through a town or countryside. The idealism of the Parkway movement in mid-century USA led to the design of high-speed roads as an art form. Designed to complement the countryside, their route is aligned to the lie of the land so that people can see the sinuous line of the road melding into the landscape and also, from a car, can enjoy the experience of seeing the countryside from of the position of a moving observer. The Cherohala Skyway, Tennessee, USA (Fig. 1.15), is a good example. The idea of a driveway intended for pleasure is a continuation of the concept of the Georgian carriage rides at Hyde Park and Regents Park, London, and incorporated into Central Park, New York, in the mid-nineteenth century, and included a verdant landscape and underpass roads. In fact all our motorways are leisure routes for at least some of their users, and even for those not on leisure trips, high-quality design and landscape are universally appreciated.

1.15 Cherohala Skyway, USA, designed as a leisure route to fit the landscape

There are also places where roads with large flows of traffic have been integrated into the urban streetscape. An obvious example is the Champs-Elysées, Paris, France (Fig. 1.16). The wide, high-capacity road is accepted there as part of the built environment.

Existing motorways are likely to be in place for a very long time, so remedial action to knit them more acceptably into the urban streetscape is being tackled over extended timescales and in association with new development. The treatment of the Coventry ring-road, virtually an urban motorway, is a good example. In order to reduce its barrier effect, it has been bridged by a new public square linking new urban renewal developments of offices on each side. These schemes require considerable resources and a long time scale.

Yet at the smaller more manageable size and the immediate to mid-term there are principles to make virtually any place that has traffic more attractive. At hundreds of locations across the country there is potential to make a local impact by improving the streetscape quality of quite ordinary streets, the sort of places that few people bother about.

1.16 Champs-Elysees, Paris, designed as a traditional street flanked by spacious pavements, landscape and buildings

Making places with traffic more attractive

Having accepted that streets are places that can be attractive, the obvious questions are how can new streets be designed to be attractive or, possibly more important, how can we improve the attractiveness of existing streets? Using a streetscape analysis, the next step is to set general objectives for improvements that are visually appropriate for the streetscape category being considered: the formality of an ordered city street or the informality of a country lane.

CONSIDER PLACE AND TRAFFIC AS A TOTAL STREETSCAPE

To convert these design objectives into action, it is necessary to understand the street as a road surface and a pavement surface, and all that happens on these as an essential part of the total streetscape. However, this is not how streets are normally understood. This might be, as we examine in Chapter 5, because our legislation, professional disciplines and even the organisation of local government tend to separate them: the design and management of roads are dealt with separately from the design and management of the places they pass through. We consider towns, villages and the countryside as pleasant places with character that simply have to put up with the nuisance of traffic, both stationary and moving. At the same time we often deal with traffic convenience and road safety as though they were totally divorced from the design of a total street. It takes a conscious effort to combine the two (Fig. 1.18).

This is how Cullen's study dealt with the urban space in the sketches of improvements to an urban streets scene, discussed in the Introduction. A traffic roundabout is transformed and incorporated into a pedestrian space that is also an elegant setting for a restored historic community building and other regeneration projects. Though many town centres have similar pedestrian-only areas, to achieve such coordinated schemes requires a long time scale, as well as extraordinary effort by a local council and/or property developer.

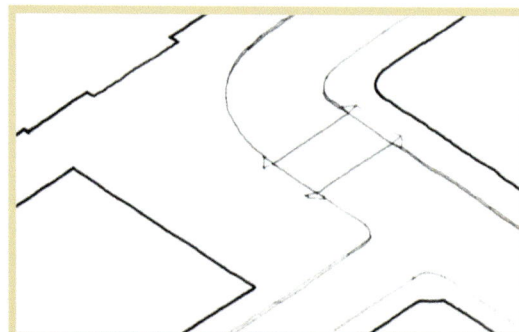

Streetscape understood as a system of roads and pavements...

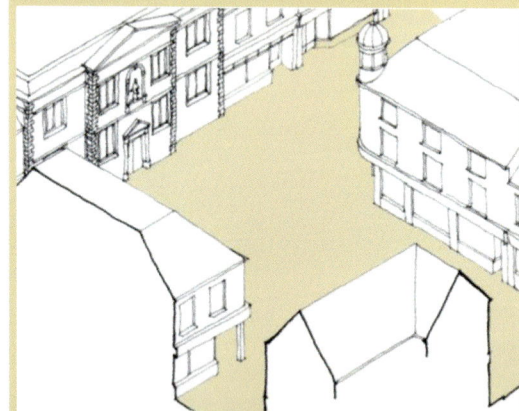

a group of buildings and a public space...

or the reality: buildings as well as roads

1.18 People see and understand a street, as recognised in the Cullen sketches, as somewhere for road traffic as well as buildings and spaces

THREE STAGES OF ACTION TO IMPROVE THE STREETSCAPE QUALITY OF ORDINARY STREETS

There are many small-scale actions that can enhance ordinary streets. Here we look again at the streetscape analysed earlier (Fig. 1.07, page 14) and apply the first two stages (Fig. 1.19) of a three-stage approach.

1.19 The first two stages of improvement to the scene at Fig. 1.07

Remove street clutter and design highway to respond to street character

1. Remove fly-posters and graffiti
2. Tidy up bins and public seating
3. Fix streetlights and signs to buildings, not posts and columns
4. Simplify traffic signs and traffic lights
5. Tidy up pavements and inspection covers
6. Simplify crossings (Road markings have been omitted in the sketch for clarity)
7. Tidy up shop fronts
8. Remove guard railings
9. Add greenery
10. Select appropriate street furniture

A reduced version of Fig 1.07 on page 14

Stage 1: Remove as much street clutter as possible.

The first of the three stages to improve streetscape quality is to remove foreground street clutter so that the positive characteristics of a place can be more fully appreciated. Because of the constant change that takes place in a street, there is relentless pressure to add more street clutter as new signs and street equipment are put in place to answer particular current needs. In time this builds up to form a permanent 'mist' of clutter that reduces the positive visual impact of a place, reducing the quality of wherever it happens to be.

The techniques to reduce street clutter are complex, because the signs or equipment that make up clutter have, or had, a practical purpose, and may have been put in place by and be the responsibility of up to a dozen separate agencies. Each may need to be dealt with separately as the purpose of the sign or equipment is re-examined using processes outlined in subsequent chapters. For example, many categories of traffic sign are not mandatory. They may be redundant and so can be removed, or the message or warning they are intended to convey can be adequately expressed through the design of local landscape. Once the clutter of non-essential signs and equipment is removed, the remainder of the equipment can be regrouped or coloured into a co-ordinated design that relates visually to the particular location.

Stage 2: Design all highway works to respond to the context or special character of the streetscape.

The second stage is to make sure that all works within and near a highway, whatever its size, scale and practical considerations, should improve the quality of the place and respond to its context. For example, if it is decided that a particular building in a street or space makes a significant contribution to the positive qualities of that place, those qualities might be enhanced and emphasised by the position and detailed design of a crossing place. A courtesy crossing could be aligned precisely to an entrance or architectural feature of a locally significant building. Where there is a hierarchy of important wide streets, less important streets and narrow urban lanes, the paving material of each could reflect a street's position in the hierarchy.

Stage 3: Where possible reorganise traffic to create new spaces.

At a larger scale or in the longer term the traffic patterns can be rearranged so that the streets and public realm spaces they pass through are enhanced or new spaces are created. This is exactly what Cullen suggests in his proposal for an urban street and is a principle being carried out in many cities and towns. Often the solution is not to totally exclude traffic but to accommodate it while respecting the local activities that take place there as well as the qualities of the place. In some streets, large volumes of traffic are unavoidable. High streets, for example, can be designed to allow pedestrians in relative comfort to walk along pleasant pavements with frequent and convenient places to cross the road. Finally, though at considerable expense, a landscape bridge can help to seamlessly integrate the activities on either side of an urban motorway that would otherwise cause intractable disruption.

Conclusion

In this chapter we have seen how streets, including roads and lanes, can be part of attractive places. Using Alexander Pope's advice that before attempting a landscape design we should "Consult the genius of the place", a streetscape analysis of the character and attractiveness of a place starts with a recognition of its category: city, town, suburb or countryside, and then goes on to examine the memorable qualities of the place within its category: the historic and prominent buildings, both as single landmarks or in groups, the spaces between buildings, including roads, and the views and vistas into and out of spaces. We have looked at past achievements to see how pleasant places have been created, and the pleasure gained from walking from one space to another.

Using the results of a streetscape analysis we have considered how to assess the potential to improve any location that incudes a street, road or lane through three stages: first by removing street clutter, secondly by designing all aspects of the street so that they respond more closely to the streetscape character of the location and, finally, where possible by reorganising traffic to create pleasant and useful public spaces. The case studies in Part 2 examine how some of these principles have been carried out in practice. In the next chapter we see how the essential needs of traffic can be accommodated in association with attractive places.

"

The way our streets are designed and managed is essential to our everyday lives and their quality affects everyone. Streets are not just a way for people to get about but are places in their own right, the centre of the community. Streets that look good can also be safer."

Robert Goodwill MP
UK Minister for Transport
(2013-2016)

"Highways and transportation professionals are increasingly aware of the wider role of roads in the community. By working closely with other disciplines and agencies we can help deliver roads that function efficiently, are safe and are memorable pleasant places to live and work."

Peter Dickinson
Chair, Urban Design Panel,
Chartered Institution of Highways
and Transportation

2.01 High-capacity dual carriageway with road-side flowers and trees. Coventry City Council

EFFICIENT MOVEMENT

CHAPTER 2

"Why don't they just get rid of all the traffic?" This is a commonly heard expression of exasperation. It is perhaps easy to wish away the nuisance of road traffic, hoping that the railway system or even canals could cope. But despite the importance of rail and water, by far the greatest volume of goods and numbers of people are moved about on roads. Even when we include travel by rail or water, a road is invariably at the beginning and end of the journey. Roads connect virtually everywhere to everywhere else (Fig. 2.01).

Road traffic has increased in proportion to the rise in GDP and is expected to continue to rise, mainly because of the increase in the numbers of journeys by private car (Fig. 2.02). Although there are exceptions, for example in central London where only about 5% of journeys to work are by car, the average for urban areas across the country is 65% and in rural areas more than 70% (Note 2.1).

Not only is traffic increasing in volume, it is becoming ever more complex. This is apparent when you look at any busy crossroads: cars, people on foot with and without disabilities, cyclists, taxis, buses and lorries all seem to be mixed together in a chaotic melee. Yet each has its own needs and expectations, and none of them can be wished away. We can't escape the need for traffic but we can manage it.

This chapter deals with the needs of modern traffic and how it is organised in urban and rural areas. It aims to unscramble what takes place at a typical road junction (Fig. 2.03), helping to identify ways in which streets can be designed to allow for efficient movement for all road users, while fitting comfortably within attractive places.

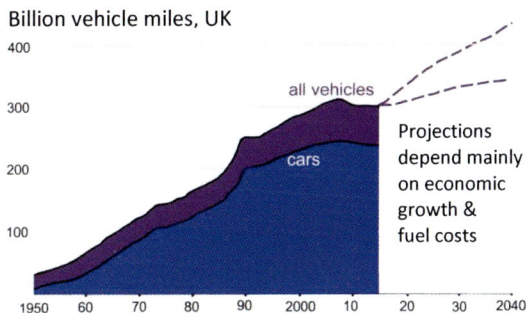

Billion vehicle miles, UK

all vehicles

cars

Projections depend mainly on economic growth & fuel costs

2.02 Private car journeys account for much of the increase in traffic – a trend that is likely to continue

2.03 A surprising number of traffic junctions look complex and ugly

Road networks and hierarchy

To increase efficiency, roads are organised into a hierarchy. At the top are roads such as motorways that are purposely designed to carry large volumes of traffic safely over long distances. At the other extreme are lanes and cul-de-sacs that mainly serve one or a few properties. The simplified hierarchy can easily be seen in the colour of the roads marked on a road atlas. Motorways are shown blue and major roads are green. Less important roads are red, and local roads yellow and white. The same road system can be understood as a network diagram, with roads shown as lines and junctions as circles (Fig. 2.04). A more refined diagram would indicate approximate volumes of traffic by the thickness of each line. Thus it is possible to easily understand the function of each road, often referred to as a link, and each junction.

With this knowledge we can think about how a junction might be improved in terms of the attractiveness of a place without compromising the efficiency of the total road network in terms of movement. A town centre junction or urban crossroad might also be an important outdoor space (Fig. 2.05). Knowledge of its function within the road network might help to produce proposals to improve the attractiveness of the place by removing, reducing or reorganising traffic flows (Chapters 7, 8 & 9).

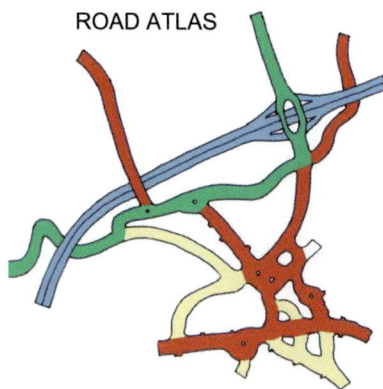

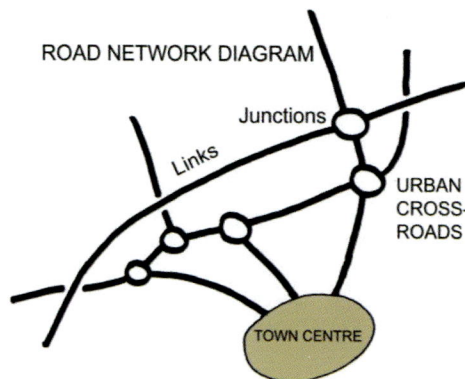

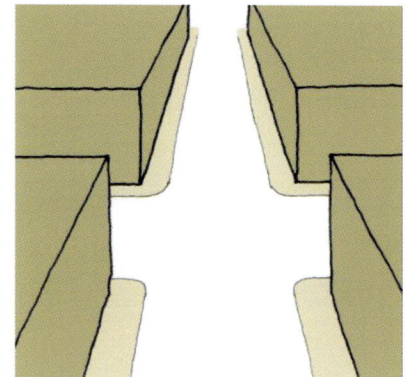

2.04 An ordinary road atlas helps us understand roads as links and junctions

2.05 An urban crossroad is a place

Road junctions

Road junctions are where most of the congestion and accidents occur, and as a result where efficient movement is impaired. At any crossroad where vehicles are able to enter and leave from all four roads there are as many as thirty-two potential conflict points between vehicle movements: when a vehicle turns right, left or simply goes straight ahead and crosses the path of other vehicles, slows to leave the flow or merges with another flow. At each of the thirty-two points in the junction vehicles could crash into, or be crashed into, from the front, back or sides. Simply by closing one of the approach roads so that there are three instead of four, the junction becomes in effect a tee. This reduces the potential conflict points from thirty-two to nine (Fig. 2.06).

That deals with vehicles, including cycles, but road junctions are also where people, including those with various disabilities, need to cross the road. Similarly it is where there is a concentration of underground service pipes and cables that need to be maintained. It is also where many prominent buildings are placed. Designing an efficient and safe, yet elegant road junction is a challenge, so ways are found to reduce the number of traffic conflict points:

1. Install traffic lights to stop one lane of traffic to allow another to pass
2. Reduce the number of directions in and out of junction
3. Help drivers to cope with the conflicts by providing a roundabout.

A fourth option, examined in more detail in this chapter, is to reduce the legal control of vehicle movements, allowing drivers to understand potential danger and make their own informed decisions on how they negotiate potential conflict points.

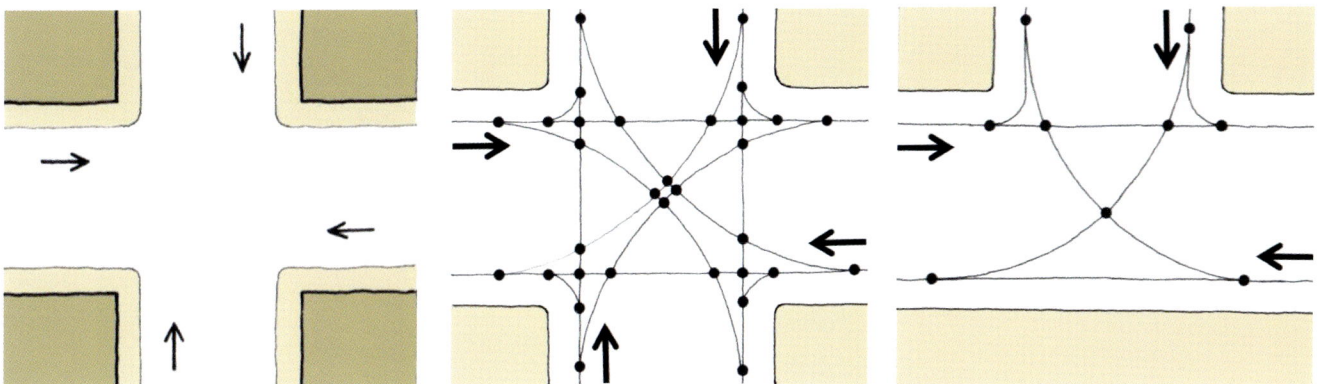

2.06 Changing a crossroad into a T junction reduces the number of traffic conflict points from thirty-two to only nine

TRAFFIC LIGHT SYSTEMS

Traffic light (traffic signal) systems are designed to be sophisticated installations that respond to the flows of traffic and pedestrians. Though they may appear at a busy junction to be remarkably complex, the principles are quite straightforward. They can be understood more readily by looking at an example of the simplest form of junction: a T junction with three roads approaching it. The traffic signals prevent or permit traffic flows so that vehicles and pedestrians move across the junction in a repeating sequence of three stages (Fig. 2.07). The first two stages control the flow of vehicles, when no pedestrians are permitted. The nine conflict points that existed without traffic signals are reduced to the one that occurs at the first stage when vehicles travelling across the junction from road C to road B cross the path of vehicles from A to C. The signal designer must judge the acceptability of this risk, bearing in mind the likelihood and time allowed in the stage for drivers to manoeuvre safely. However, many signal installations include similar stages that require drivers to make their own judgments about when it is safe to turn across the path of, or join, another lane of traffic. Traffic-signalled junctions are therefore rarely totally free of risk.

DESIGN OF SIGNAL HEADS AND BRACKETS

In addition to the number of signal heads needed, a matter that affects the quality of a streetscape is their appearance. An elegant arrangement is examined in the Kensington High Street case study in Chapter 8.

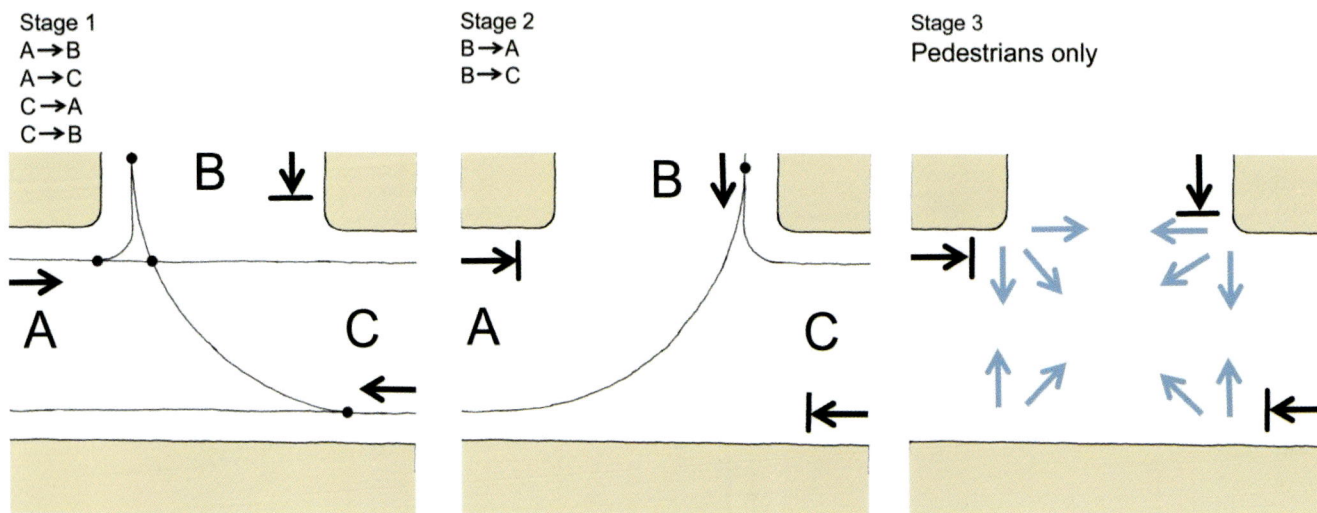

Stage 1
A→B
A→C
C→A
C→B

Stage 2
B→A
B→C

Stage 3
Pedestrians only

2.07 At the third stage of this three stage traffic light (traffic signal) sequence people can cross the roads in any direction they wish

ONE-WAY STREETS

In order to improve the efficient movement of vehicles and maximise the use of restricted town-centre road space, many town-centre road junctions have been simplified and made part of a wider one-way traffic system. By permitting vehicles to leave the junction from only two roads instead of all three, the number of potential conflict points at a T junction is reduced from nine to four. Installing traffic lights can further reduce the number of conflict points from four to none (Fig. 2.08). The diagram also shows a typical arrangement where pedestrians are signalled to cross the road at a place and time that can be fitted into the sequence of traffic light stages that control the movement of the vehicles. Though efficient for traffic, pedestrians are inconvenienced. They are seldom able to walk straight across the whole road without having to stop in the middle of the road and wait for the next phase of the traffic lights. The wait is inconvenient, uncomfortable and at a place where they also may feel vulnerable on an oddly shaped area of pavement not needed for moving vehicles.

The feeling of being second-class citizens is emphasised by pedestrians being required to walk round unpleasant and poorly maintained guard railing (Fig. 2.09). The need for guard railing is examined more fully in the Kensington High Street case study in Chapter 8. This was one of the first schemes in the UK to challenge the effectiveness of guard railing and has led to changing attitudes on its need as well as further studies of the road safety evidence that underpinned its provision (Note 2.2).

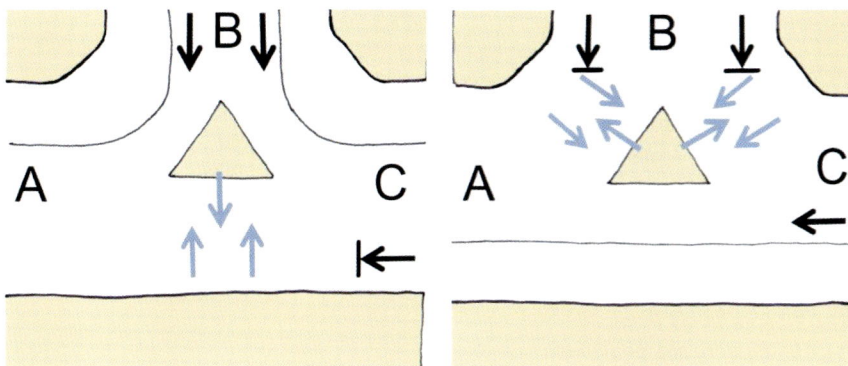

2.08 One-way streets and traffic lights can remove all traffic conflict points

2.09 Guard railing seldom improves road safety

ROUNDABOUTS

Roundabouts are intended to reduce conflicts in a different way to traffic lights. Drivers enter or leave the roundabout by a protocol of either giving way to vehicles already on the roundabout or being given way to by vehicles entering the roundabout. Roundabouts allow large volumes of traffic to manoeuvre efficiently and are less costly to maintain than traffic signals. When signing is reduced to the legal minimum and landscape is added, they can be quite elegant. There are disadvantages as they need more land than a traffic signal system and so are difficult to fit into an existing urban location without radically changing the character of the streetscape. The main disadvantage is that large roundabouts cannot, without some modification, allow pedestrians and cyclists to use them in comfort and safety (Fig. 2.10), and the modifications can appear quite unattractive.

Mini-roundabouts, as their name implies, are smaller versions. However though mini-roundabouts are safe and convenient for drivers, they are remarkably inconvenient for pedestrians, especially those with disabilities, who need to cross the road at the junction (Fig. 2.11).

The legal requirements for signing, including traffic signals, at junctions and crossings where traffic is controlled are set out in the national traffic signs regulations (Note 2.3). The attractiveness of a place can certainly be improved by reducing signing to the legal minimum. It is possible to go further and dispense with signing almost completely at uncontrolled junctions.

UNCONTROLLED JUNCTIONS

The need for the strict control of traffic is obvious with respect to motorways. Speed and volumes of traffic are high and efficient movement relies on drivers being clearly instructed on what to do.

In some cases traffic at local road junctions and crossings is not legally controlled. The design of the road layout and its immediate surroundings encourage drivers to act responsibly and drive safely. Examples at Gosford Street, Coventry and Fountain Place, Poynton are examined at Chapters 7 and 9). A suburban example at Longmoor Street, Poundbury is studied at Chapter 10. The advantages in streetscape terms are that such streets are efficient for traffic movement without the need for additional traffic signs or equipment. The safety aspects are dealt with in the next chapter.

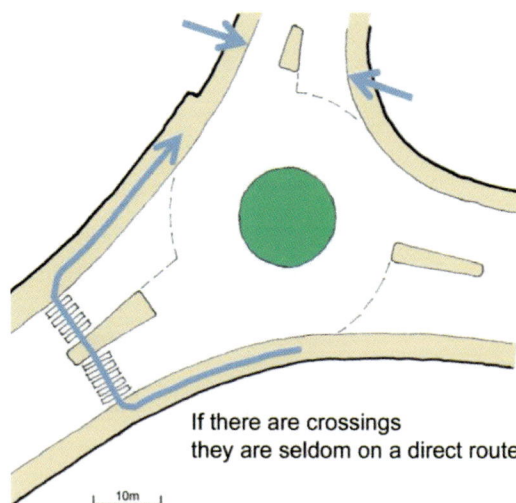

If there are crossings
they are seldom on a direct route

10m

2.10 Conventional roundabouts efficiently handle large volumes of traffic but can be inconvenient for pedestrians because controlled crossings (zebra, pelican, etc.) are usually positioned away from a direct pedestrian route (Chapter 7)

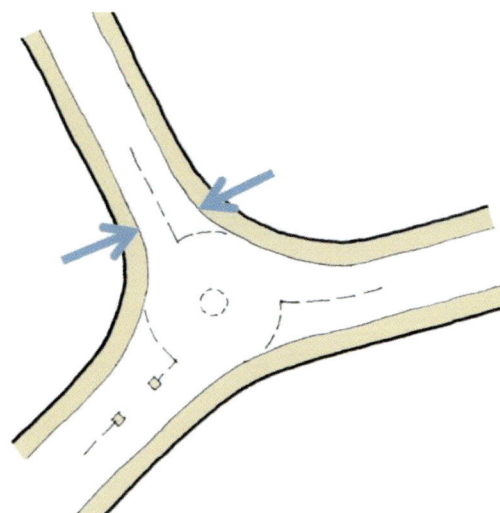

2.11 Mini-roundabouts can also be difficult for pedestrians, particularly those with disabilities

Pedestrian crossings

What becomes very clear in any discussion about the design of civilised streets is the importance of balancing the requirements for the efficient movement of traffic with the needs of pedestrians. Though a completely segregated system of underpasses or bridges over a road for pedestrians seems sensible, it is rarely convenient for pedestrians as it usually involves a detour or a walk down dingy underground passages. Alternatively, to separate vehicles from pedestrians by physical barriers such as railings is difficult to fit attractively into a streetscape and would still need to include at some point a place where people cross the road.

The main categories of crossing are signal-controlled, whether at a junction or stand-alone, where drivers are required to stop at a red signal, zebra crossings where drivers are required to give way to pedestrians on the crossing, and courtesy crossings where drivers stop voluntarily.

SIGNAL-CONTROLLED CROSSINGS

In many respects signal-controlled crossings are similar to the pedestrian crossing places within a traffic signal system at a road junction. They provide time for traffic movement at places where the heavy use by pedestrians at a zebra crossing would impede traffic flow. In addition many people, particularly those with disabilities, appreciate the security of the legal enforcement and the degree of certainty that vehicles will stop at the signal. However, they are not foolproof. Typically accidents occur when able pedestrians become frustrated at having to wait, and take risks.

2.12 This urban signal-controlled crossing at a junction deals with traffic movement efficiently, helps people cross the road comfortably and is quite clutter free

In order to reduce vehicle flows as little as possible, some crossings stop traffic in each direction independently so that pedestrians need to wait twice before crossing the road completely, often waiting in the middle of the road 'sheep pen'. More elegant arrangements are discussed in the case studies (Chapters 6 and 8). However, the zigzag road markings are a legal requirement, indicating the ban on stopping (except for complying with the signals) or overtaking.

ZEBRA CROSSINGS

Zebra crossings with minor variations are known across the world. In the UK we have the ubiquitous Belisha beacon with its distinctive yellow illuminated globe on a black-and-white striped post.

From a streetscape viewpoint there is a reservation that they need to be noticed by drivers and have an easily identified appearance that is common across the country and that stands out in the street scene: the very appearance that sometimes make them difficult to fit into the subtle individual visual character of a place. The national design rules are clearly stated in the traffic sign regulations (Note 2.3). These include the precise dimensions and location of the post, with its globe and black-and-white bands, as well as the black-and-white stripes on the road and rows of adjacent white zigzag lines. It is, however, possible to vary their width from the standard 2.4 metres up to a maximum of 10 metres to form part of a landscape setting of a prominent building (Figs. 2.13 & 2.14).

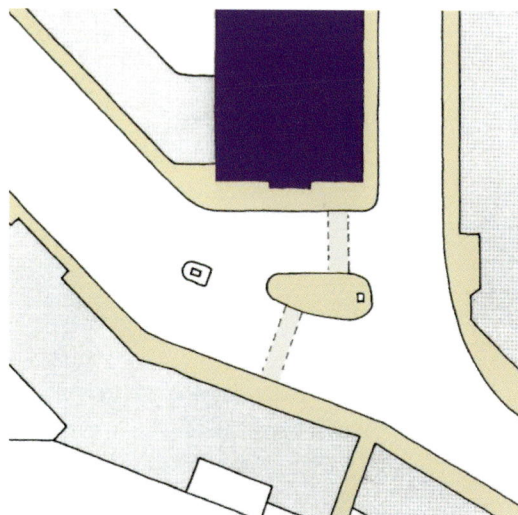

2.13 A wider zebra crossing at Drury Lane leads to Great Queen Street, London

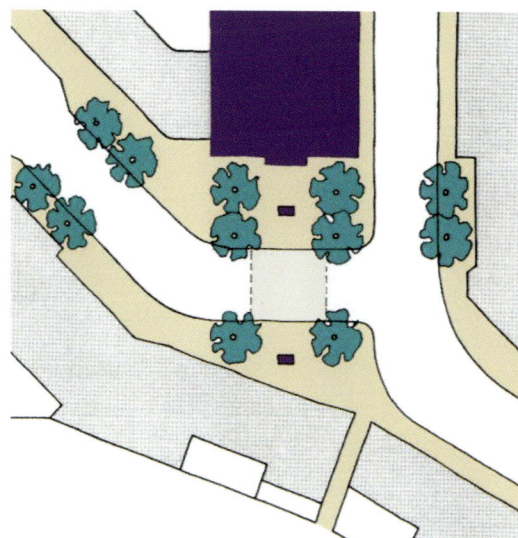

2.14 Crossings can be part of the wider landscape

2.15 Abbey Gate, Bury St Edmunds, Suffolk

2.16 The Corn Exchange, Devizes, Wiltshire

2.17 High Holborn, London

COURTESY CROSSINGS

As their name implies courtesy crossings are places where, as a courtesy, drivers stop to allow pedestrians to cross the road. There is no legal requirement for drivers to do so and traffic is not controlled, so they are also known as uncontrolled crossings. Drivers are more likely to want to stop if they can see and understand the reason, are travelling at a speed that allows them to stop and are not too inconvenienced if they do stop.

Well-designed courtesy crossings are shown to encourage more drivers to stop for pedestrians (known as the yield factor) than at some zebra crossings. The feel of the street: its landscape, the position and relationship of buildings to the road, as well as the road surface, all have an effect on driver behaviour. There are sufficient variations and adaptations of courtesy crossings for it to be possible more often to locate them exactly where pedestrians wish to cross a road, rather than, as at signal-controlled and zebra crossings, where it is convenient for drivers to stop. Providing the approach speeds are low, say about 20mph, there is a wide range of possible design variations that can fit the character of a place and often enhance its attractiveness. Six examples are:

1. A two-stage courtesy crossing that is easier for pedestrians to safely use is located on a direct pedestrian desire line and aligned with the entrance to a historic building visitor attraction (Fig. 2.15)

2. At another location a central pedestrian refuge is aligned with the central doors of a community hall (Fig. 2.16)

3. A simply designed crossing over a two-way road is indicated only by a coloured road surface at the edge of an uncontrolled junction near the university buildings of Coventry (Chapter 7)

4. A two-stage crossing at the entrance to a road junction, designed to allow pedestrians to cross exactly where they would wish, as at Poynton, illustrates that the principles could be applied at many mini-roundabouts (Chapter 9)

5. Courtesy crossings at intervals along a street, across the entrances to side roads and squares at Poundbury (Chapter 10)

6. Keep Clear road markings indicate a place for people to cross a city centre road in a direct line between two pedestrian passages but some fifty metres from a signal-controlled crossing. A less innovative design might have installed railings to force people to use the signal-controlled crossing (Fig. 2.17).

Traffic calming

Traffic calming refers to physical measures that force or encourage drivers to drive slowly (Note 2.4). Speed humps are a common example but here we consider alternatives that fit more pleasantly into the streetscape. First is an example based on the principles being applied at the entrance of a small town or village at the point where a 30mph speed limit begins. Instead of a festoon of warning signs, the minimum signs that are required by law are put in place at each side of the road, with a road narrowing and the removal of the centre-of-road white line. This, together with landscape, reminds drivers to reduce speed and sets a standard of design for all the traffic calming throughout the village (Fig. 2.18). The principle is to design traffic calming devices that are not simply a collection of traffic warning signs, bollards, kerbs and traffic management equipment (Fig. 2.19) but achieve their object to influence driver behaviour, and fit into or enhance the streetscape character.

A few alternative ideas for innovative design are first an arrangement of parked cars set at right-angles to the road, between trees in the road and flower beds. It has existed on the A509 in the village of Olney for several years. Drivers travelling along the street look out for and stop to allow parked cars to be driven out into the road (Fig. 2.20).

Using this theme, and combining it with courtesy crossings and car parking places, gives very wide scope to design traffic calming schemes that fit exactly the character of a location and actually enhance it. The aim would be for the final scheme to be blended completely with its context so that any intervention in the scene to deal with traffic is not apparent. Some examples are:

1. Physical road narrowing with crossing and street trees (Fig. 2.21)
2. Road narrowed with wider verges, trees and parked cars, possibly incorporating a courtesy crossing (Fig. 2.22)
3. Road narrowed simply by arranging places where cars are intended to park (Fig. 2.23).

In each case drivers will need warning that the road is narrowed, but that can be by landscape and road texture, rather than merely signs and white lines. Completed examples of similar techniques are examined in the case studies at Chapters 9, 10 & 11.

2.18 Clifton, Cumbria. White line removed on A6

2.19 Less attractive bollards, posts and signs

2.20 Right angled parking calms traffic on A509

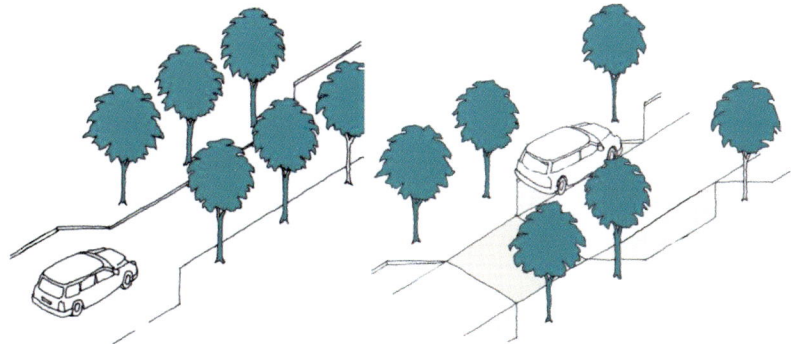

2.21 Physical road narrowing with crossing and street trees

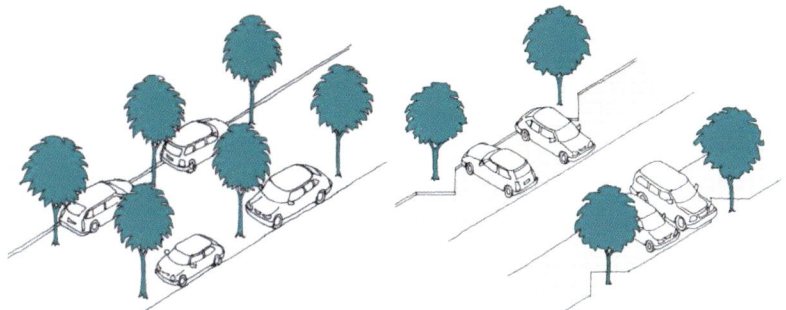

2.22 Road narrowed with wide verge, trees in road or car parking

Box 2.1: Chart showing recommended distance between of speed reducing measures

Danish planning and design procedures for traffic calming. Danish Road Directorate

Distance between Measures (m)	Reference speed (km/h)
25 (max 50)	10 – 20
75	30
100	40
150	50

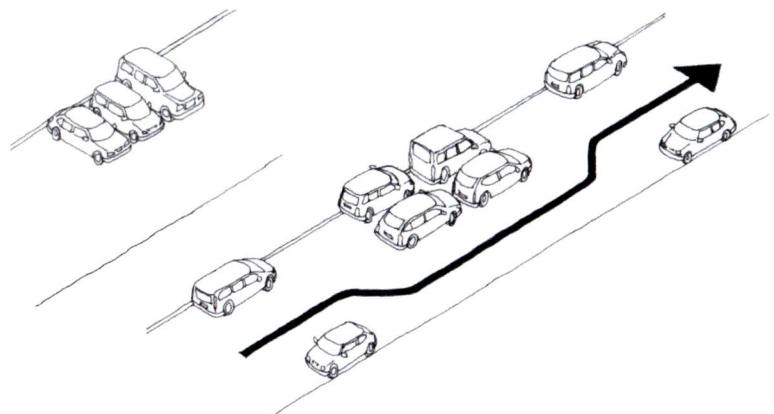

2.23 Road narrowed by the position of car parking places

HOME ZONES AND QUIET LANES

The concept of a Home Zone, a residential area where drivers can no longer assume that other road users will give way to them, became established in UK transport legislation with the introduction of Home Zones and Quiet Lanes in 2006 (Note 2.5). A target speed of no more than 20mph is achieved through traffic calming measures, integrated into the design rather than as an engineering afterthought. Possibly because they are intended for purely residential areas, Home Zones have been applied mostly to the centres of residential areas and new housing schemes.

2.24 Lorry turning areas need not be defined by concrete kerbs. They can be fitted into landscape

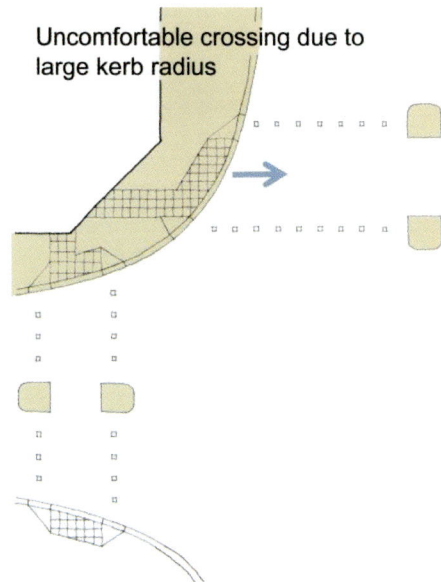

Uncomfortable crossing due to large kerb radius

2.25 Large kerb radius allows lorries to turn easily but is uncomfortable for pedestrians

Road geometry

The primary purpose of the layout or geometry of roads is to facilitate the efficient movement of vehicles, particularly large vehicles when they need to turn. Instead of building ugly areas of road in the form of a Y, T or 'hammer head', it is possible to incorporate the necessary area of hard ground surface into a wider landscaped road scheme such as a square (Fig. 2.24). An example is at Burraton Square Poundbury (Chapter 10). There are also places where occasional waiting vehicles can be accommodated in a road without the need for over-elaborate infrastructure. An example is at Fountain Place, Poynton, where vehicles waiting at the St George's churchyard gates stop on the road junction (Chapter 9).

Another common problem occurs at many urban street corners where a large kerb radius is provided to allow long vehicles to turn. This is uncomfortable for pedestrians, especially disabled people (Fig. 2.25). A more elegant alternative design keeps a tight kerb radius that retains a straight kerb that is more comfortable for pedestrians, but ensures that the total road width is sufficient to accommodate the track of a long vehicle's wheels as it turns (Figs. 2.26 & 2.27).

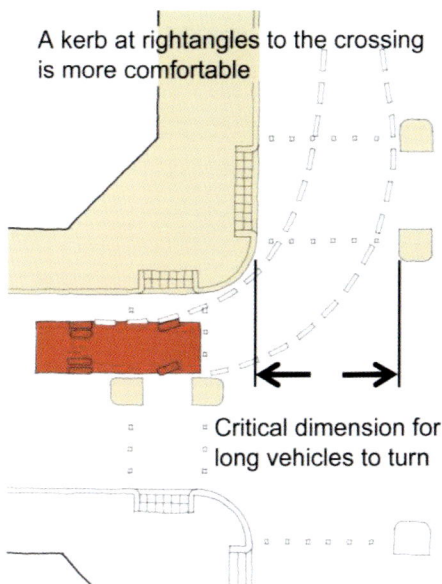

A kerb at rightangles to the crossing is more comfortable

Critical dimension for long vehicles to turn

2.26 An alternative tight kerb radius with sufficient road width is better for pedestrians

2.27 A tight kerb radius at Sloane Street, London, is typical of the Georgian and Victorian street layouts in many historic town centres

Less regulated traffic

We have seen how the design of uncontrolled road junctions, courtesy crossings and traffic calming is developing in ways that could improve the attractiveness of a streetscape. Landscape and the character of a streetscape, instead of traffic regulations and road signs, are used to influence driver behaviour. This relies on a greater understanding of how a driver perceives a road, considered in more detail in the next chapter on road safety. Also, as we consider in Chapter 4, streets now need to cope with far more types of user than they were initially intended. The view is growing among some professional designers that when traffic is less regulated, particularly in mixed-use streets, and drivers are able to take more responsibility for their actions, the result is efficient, safe and more pleasant streets. Three examples from outside the UK give a flavour of how other countries deal with similar issues: an all-way stop, pedestrian priorities at crossroads and the Swiss 'Encounter Zones'.

ALL-WAY STOP

Used extensively in the USA, particularly at rural crossroad locations, the principle of an all-way stop is that at some crossroads there is no need for either traffic signals or a roundabout. With only a stop sign at each approach road, drivers understand that the first to arrive at the junction has priority and proceeds across it. At Durban, South Africa, for example, the principle has been adopted at a city centre crossroads, each with two lanes of approaching traffic and pedestrian crossings. Though the rules in the USA are a little different, it demonstrates that when drivers of a nation have become familiar with an alternative system, it can be safe. The advantage to the streetscape is that the clutter of traffic signals or the land required for roundabouts are not needed.

PEDESTRIAN PRIORITY AT CROSSROADS

A feature of signal installations in the USA compared with those in the UK is that in the States a single fitting suspended above a crossroad appears to be sufficient, while in the UK a similar crossroad might need a dozen signal heads, each mounted on its own post. One reason is that drivers turning at a junction in the USA are required to yield to pedestrians. This removes the necessity for separate signals to control turning traffic.

SWISS 'ENCOUNTER ZONES'

Many European countries have zones where drivers are expected to give way to pedestrians. A Swiss 'Encounter (or meeting) Zone' is similar to the concept of a UK Home Zone but with three important additional features:

1. A legally enforced speed limit of 12mph with the necessity for drivers to yield to pedestrians is enforced by legislation
2. A zone may be located in a commercial as well as a residential area
3. Parking is only permitted at marked locations so this, in addition to there being less need for traffic regulation signs within the zone, results in a quite uncluttered streetscape.

Conclusion

At the conclusion of this chapter we have a clear idea that the way traffic is organised to move efficiently affects the quality of streetscape, and that every road and junction fulfils a function as part of a highway network. The greatest challenges and opportunities for streetscape improvement occur at road junctions and where people cross the road. The efficient movement of traffic cannot be wished away; it is essential for virtually everything we depend on for daily life. However, with a greater understanding of how drivers cope at traffic signals and roundabouts, and how pedestrians cross the road at variously designed crossings, we can appreciate the many options for improving the ordinary everyday street. We also have demonstrated that each category of intervention or work to a street can be designed to enhance an area and respond to its context in order to make it more attractive.

Current thinking is that alternative designs that give drivers greater responsibility, such as at uncontrolled junctions and courtesy crossings, also have the benefit of enhancing streetscape. This can be done while maintaining the efficient movement of traffic. But the questions remain: are these measures safe, and are they lawful? We deal with both these issues in the next chapter.

"

By working together across disciplines in partnership and by building coalitions, far more can be achieved than by trying to go it alone. Secondly for a road safety innovation to be successful it must be based on sound evidence. The need for it must be clear and its effect must be rigorously evaluated."

HRH Prince Michael of Kent
Patron of the Commission for Global Road Safety

"There are many aspects to road safety. Much depends upon the abilities and reaction of road users when an unanticipated event occurs. Highway Engineers help to create legible streets and pleasant places that operate efficiently for movement, and minimise the likelihood of driver errors and collisions."

Tony Kirby
President, Institute of Highway Engineers

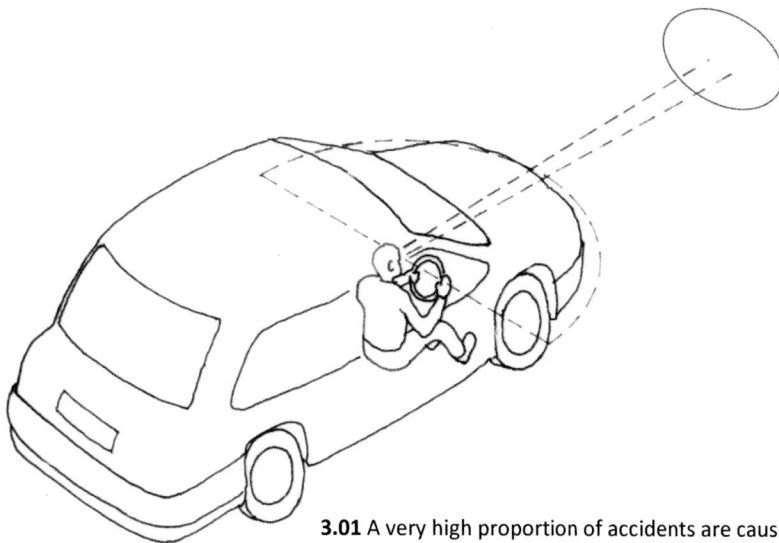

3.01 A very high proportion of accidents are caused by driver error, often through lapses in concentration

ROAD SAFETY

CHAPTER 3

The number of road accidents that lead to death or serious injury is frighteningly high, often tragically due to driver error (Fig 3.01). The annual total, 24,000 (Note 3.1), equates to several airliner crashes a month, yet very few accidents are reported nationally. In comparison, there were no accidents in 2015 on commercial airlines operating to or from the UK (Note 3.2).

Fortunately, until 2010 there was a gradual decline in the number of road accidents (Fig. 3.02), but this is probably because the vehicles themselves became safer as brakes, seat-belts, crumple zones, vehicle structure, as well as the speed and quality of medical care improved. Since 2010 the numbers have remained quite constant, with any variations within statistical limits (Note 3.3). Accidents involving pedestrians have not significantly reduced.

Accidents involving child pedestrians vary according to locality, probably because in many places children are simply not allowed out on their own, so are less likely to be in a road accident. Accidents involving cyclists, though few compared to other road users, are in proportion to distance travelled very high and unfortunately are increasing, especially at weekends.

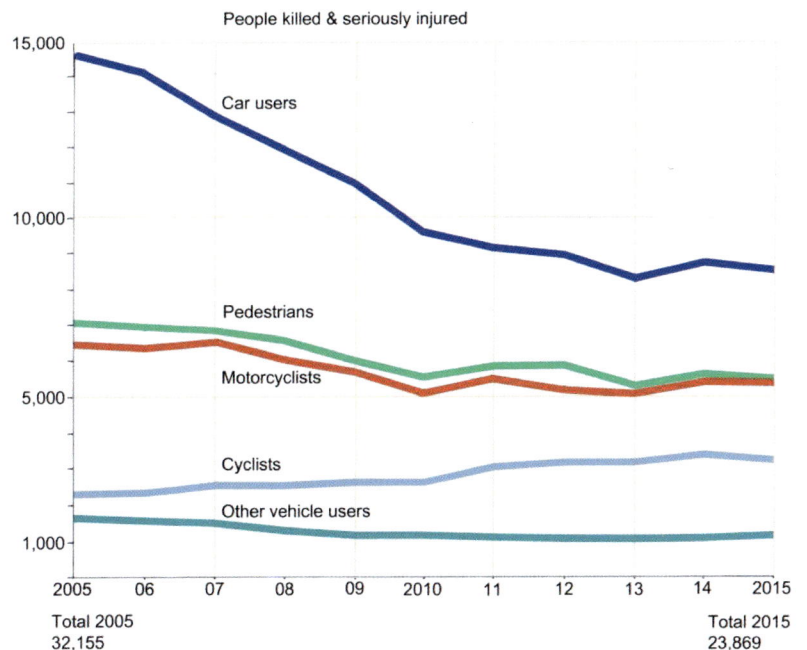

People killed & seriously injured

Car users

Pedestrians

Motorcyclists

Cyclists

Other vehicle users

Total 2005
32,155

Total 2015
23,869

3.02 The number of road accidents is frighteningly high, and between 2010 and 2015 (and into 2017) have remained statistically quite constant

Of course, statistics need to be interpreted with caution. As a background note journeys by road tend to increase as a proportion of the nation's gross domestic product, GDP, and the number of road accidents follows the volume of traffic flows to the extent that there are peaks and troughs throughout the week and during each day. The graphs of the number and time of accidents clearly show that on a typical Friday accidents involving both car occupants and cyclists follow the fluctuating traffic flows of the working day (Fig. 3.03). On Sundays the number of accidents involving cyclists peaks at 11.00am while those involving people in cars peak at 1.00pm. At about 8.00am on Fridays and at 11.00am on Sundays the accidents involving cyclists are at a similar level to those involving car users. This is remarkable as the average number of miles travelled in a year by cyclists is only one per cent of the average number of miles travelled by people in cars. The conclusion is that cycling is disproportionately dangerous.

Against the background of these statistics it is understandable that many of the on-going adjustments to the design and maintenance of a street are the result of concerns about road safety. The two previous chapters have examined how the arrangements for efficient traffic movement can be incorporated into and help make places attractive. We now look at the legitimate concerns about road safety and the causes of accidents. The chapter then explains current thinking on the application of our knowledge of driver behaviour to the design of streets that are attractive, efficient and help people to travel safely.

People killed and seriously injured during an average day

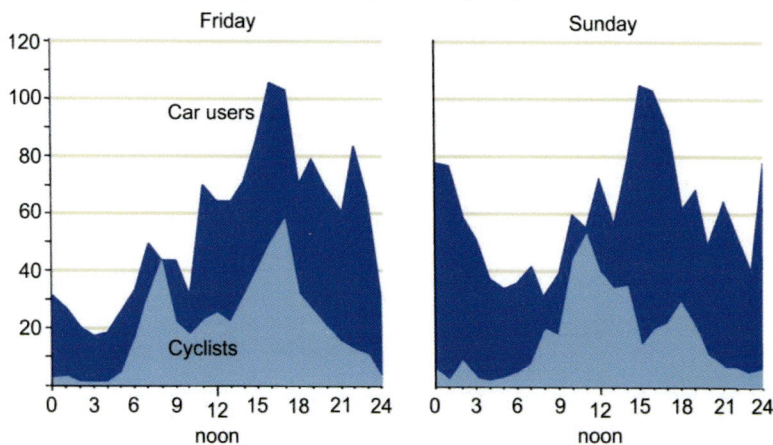

3.03 The number of cyclist casualties compared with car user casualties is remarkably high. Cyclists travel about one per cent of the total distance of car users

How accidents happen – driver behaviour

Most accidents happen when someone makes a mistake or an error of judgment during the difficult and complex task of driving or using a road. An error is very seldom the result of a single factor or occurrence, but it usually involves driver behaviour – a term that describes the ability of a driver to:

1. Expect a potential hazard

2. See and understand it

3. React to it and drive safely.

To appreciate how drivers see a potential hazard requires an acknowledgement of the limitations of the human eye. For example, after an accident drivers may say they did not see something. That could be true. It is because our eyes have quite severe physical limitations in terms of what they can see. Secondly, drivers may also say they did not see something when they mean that they did not notice it or fully understand the implications of it. This is often termed inattentional blindness. Again this is very common as our brains are easily distracted from the complex task of driving. Thirdly, even if drivers can see and understand a potential hazard they need to have the ability to react safely, and if necessary have the time and space to slow or stop.

The design of motorways illustrates the amount of thought and expenditure that is given to road safety. All the conditions to help drivers are in place: the road ahead is easy to see and to understand; speed is strictly controlled; turning and opposing streams of traffic are protected from each other; the categories of road users are limited; and if it is necessary for drivers to stop, they are given clear instructions on how to do so within an adequate time and distance. The result is that motorways are by far the safest category of road for drivers to use. However, the remaining roads in the network, that is the majority, are more complex for drivers than is generally recognised. This is why an understanding of driver behaviour is helpful when considering street design, starting with the limitations of human sight.

WHAT A DRIVER ACTUALLY SEES AND UNDERSTANDS

Over hundreds of thousands of years human beings have evolved to hunt and to run at a speed of not more than 20 miles an hour. We were not intended to drive cars at 70mph. In fact our eyes and brains are particularly suited for hunting. The eye sees in very fine accurate focus a very narrow cone of vision of little more than two degrees. At the same time we are able to be aware of any movement or large objects in a very wide cone of almost 180 degrees (Fig. 3.04). For hunting this is superb. We can focus on a particular juicy quarry at some distance while being aware that we ourselves are not the quarry of some larger unpleasant predator.

However, this adaptation of eyesight is not ideal for the modern activity of driving. Only being able to see a narrow cone of vision in detailed focus, and only one at a time, means that in order to understand what is happening in the whole scene, we have to constantly move our eyes (Fig. 3.05). This we do at the rate of about three or four eye movements a second.

The limitation of the narrow focus and only three eye movements a second is compounded if drivers need to concentrate and dwell only on one object either in order to try to understand what it is or if they find it particularly interesting. But at the same time the vehicle they are driving is continuing along a road. Three eye movements a second, at the legal urban area speed limit of 30mph, equate to one eye movement every 13.5 metres. If drivers look away from the road ahead for two seconds, they could simply not see anything else on the road for a distance of 80 metres. That is sufficient to fail to see a child or cyclist on the road.

Even if drivers are normally able to physically see something on the road there are many factors that may prevent them noticing it. Usually, to compensate for the considerable limitations of the eye, the brain works closely with the eyes, fills in the gaps, based on past experience and makes assumptions so that the driver can usually correctly understand what is ahead. In addition, because there is so much to look at and no time to focus on everything, experienced drivers become very good at looking for the right things. Unfortunately, people also tend not to see something that they are not actually looking for, even if it is large and brightly coloured, or possibly a cyclist (Fig. 3.06).

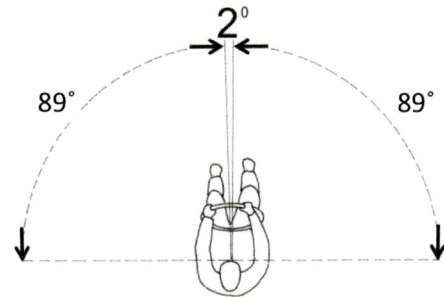

3.04 The human eye can only see a two-degree cone in sharp focus at one time

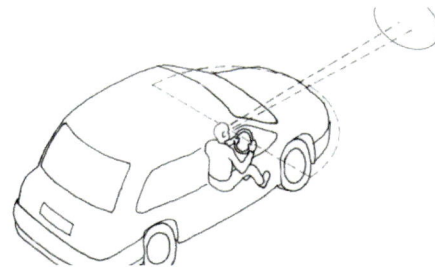

3.05 Drivers move their eyes three or four times a second to try to understand what is ahead

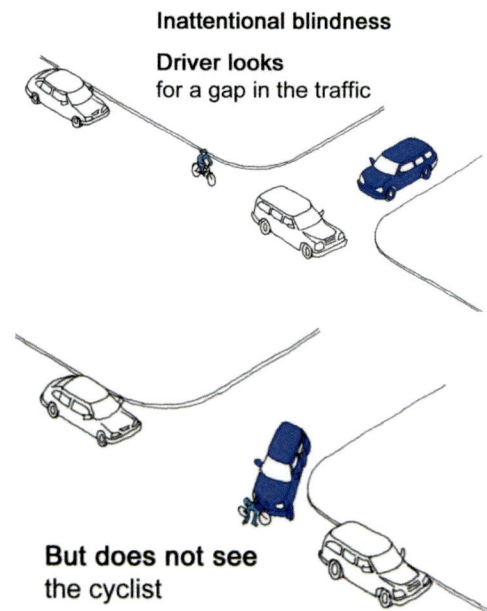

Inattentional blindness

Driver looks for a gap in the traffic

But does not see the cyclist

3.06 Drivers may not see things they do not look for

3.08 Signs intended to be helpful may be a distraction. This sign has too much information for drivers to understand in time

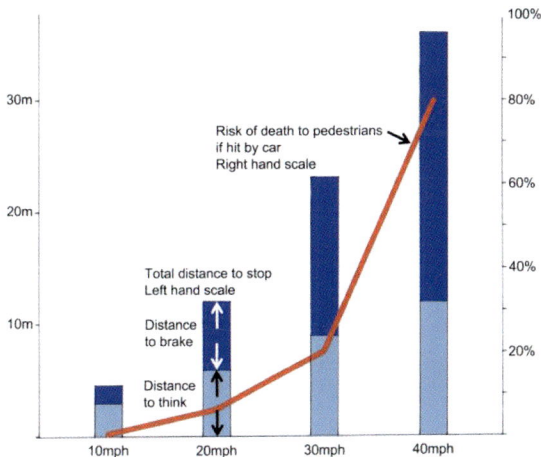

3.09 Both the likelihood and severity of accidents increase disproportionately to an increase in speed

DISTRACTION AND MULTI-TASKING AFFECT UNDERSTANDING

In addition to the physical limitations of drivers' eyesight and the problem of inattentional blindness, a high proportion of accidents are caused by drivers being distracted (Note 3.4). Some studies find that distraction caused, for example, by preparing to eat, eating, drinking, spilling a drink or using a mobile phone (even if hands free) while driving, is a contributory factor in the cause of crashes. Too many signs, or sign overload, can also distract drivers as they either take time to distinguish between important and irrelevant information or take their eyes off the road ahead completely (Fig. 3.08).

DRIVERS' ABILITY TO REACT IN TIME: THE EFFECTS OF SPEED

Assuming drivers are able to see and understand a hazard, they still need to be able to react in time and drive safely. The distance a vehicle needs to stop depends on the time taken by a driver to react and apply the brakes, and the speed it is travelling. Drivers respond less quickly to an unexpected hazard than to one that is expected, and their reaction times are reduced by alcohol and tiredness, and can vary with age. Quick reactions are particularly important at higher speeds. Although the distance needed to think increases in proportion to an increase in speed (Fig 3.09 left hand scale), the distance needed to stop once the brakes are applied will be quadrupled: from an average of 6 metres at 20mph to 24 metres at 40mph (Note 3.5). A higher speed therefore reduces the chance that drivers will notice a potential hazard and will be able to react in time, and so increases the likelihood of an accident.

In addition to increasing the likelihood of an accident, a higher speed also disproportionately increases the severity of any accident that does occur. Newton's law of force and deceleration is borne out by everyday experience of the vulnerability of road users. People in a car hit by a lorry will come off much worse than the occupants of the lorry. Walkers and cyclists hit by a vehicle will suffer more if the vehicle is travelling quickly. The chance of a pedestrian being killed or seriously injured if struck by a vehicle travelling at 20mph is less than 3%. At 30mph, the likelihood increases to 20% and at 40mph, to 90% (Fig.3.09 right hand scale).

Improving road safety

In a book dealing primarily with aspects of streetscape we seek to identify how to respect and enhance the attractiveness of a place and maintain efficient movement, but also to create conditions where drivers are less likely to make errors of judgment and cause accidents. This is examined more fully in Part II, where each of the case studies demonstrates a form of self-explaining road.

SELF-EXPLAINING ROADS

The theory of self-explaining roads offers a supplementary approach to the design of safe roads. It recognises that without specific instructions and signs, drivers are conscious of the environment they are driving through, such as when they turn from a main road into a town and change their driving behaviour accordingly. Conventional arrangements of traffic signals or roundabouts at road junctions such as those we considered in Chapter 2 are not totally accident free. Despite the number of signals and restrictions, the highest proportion of accidents in urban areas occur at road junctions and along main roads. One reason is that when drivers are shown a green light they act as though they have total ownership of the road. When they do occur, accidents at signalled road junctions are therefore likely to be severe, because a vehicle would probably be travelling at speed and the driver would not expect any sort of hazard.

For the same reason guard railings give drivers a false sense of certainty that there will be nothing in the road ahead and that they can proceed at speed. Similarly, a road may give drivers an incorrect sense of security by appearing to be safer than it really is. For example, a main road may appear to have all the built-in safety features of a motorway, such as spacious entrances and exits to allow time for drivers to negotiate, and restrictions on the types of users, when in fact it has frequent potentially hazardous junctions and a variety of users, including slow-moving, vulnerable cyclists.

A self-explaining road is one that is designed so that a driver knows what to expect in the road ahead and is able to react safely. It acknowledges the principle that drivers are influenced by their surroundings and, given the right conditions, can take greater responsibility for their own decisions and safety, and the safety of others. This proposition is supported by comparing the accident records at two adjacent locations in north London. The first is a short but seemingly dangerous stretch of road at the Spaniards Toll Gate and public house, Hampstead, London (Fig. 3.10), where the road narrows, turns a blind corner and changes level. The second, a short distance away on the same road and therefore carrying much the same volume of traffic, is at a spacious T junction with Winnington Road (Fig. 3.11). It has all the benefits conventionally associated with safety: a wide road, good sight lines, illuminated bollards, generous proportions and spaciousness at the junction. However, the published accident data show the T junction to have far more accidents.

The conclusion is that drivers recognise the possibility of a potential hazard at the public house and proceed with more care. Other experimental schemes have demonstrated similar results, such as traffic speed being reduced where centre-of-road white lines have been removed, where there is a change of level or texture of road surface, and where a traffic lane is designed to appear narrower and more dangerous than it is (Chapter 9).

3.10 The seemingly dangerous pinch point where the road changes direction and level has a good safety record: one slight accident in 10 years (source Crashmap). Spaniards Toll Gate, Hampstead, London

3.11 The seemingly safe, spacious adjacent T junction has a poor safety record: 16 slight and two serious accidents in 10 years (Crashmap). Hampstead Lane – Winnington Road, London

Because drivers react more quickly when they expect a hazard than if they do not, the first requirement of a self-explaining road is that drivers should be aware of the possibility of a hazard, without the need for excessive traffic signs or regulation, then see and understand a hazard and react safely. In principle, in a self-explaining road it is possible and practical for a driver to:

1. Expect a potential hazard
2. See and understand it
3. React in time and drive safely.

1. Drivers need to expect the possibility of a potential hazard

This road safety principle chimes with the streetscape objective to emphasise local character. Drivers take notice of where they are driving, and they respond to the total character and atmosphere of the street to the extent that it can be designed to give drivers subliminal messages about how they should drive. What happens on the pavement adjacent to the road helps to indicate to drivers what to expect. For example, shops, markets and other places where people congregate indicate that people will be near the road. Similarly, instead of a sign indicating a school, images or sculpture at the roadside can give a powerful message, and artefacts that suggest local community activities help give the impression that a village street is part of a real community and that drivers should take greater care there. In each case drivers are made less certain that the road ahead is clear and will reduce their speed in order to cope with a possible potential hazard, even if it is people walking in the road (Fig. 3.12).

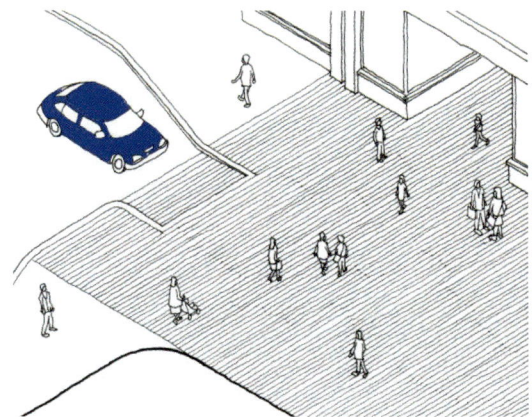

3.12 Drivers need to expect the possibility of a potential hazard

2. Drivers need to see and understand a hazard

Having reduced speed and being aware of the possibility of a hazard, drivers need to be able to see and understand the nature of the hazard. In practical terms that would mean negotiating with other drivers at a junction as they would at a roundabout or being able to see clearly if there was a cyclist or pedestrian, possibly with a disability, on or about to go on to the road ahead. At this point there should be as few distractions as possible. It helps if courtesy crossings, for example, span a two-way road in two distinct stages, so drivers can concentrate solely on and yield to the pedestrians on the crossing and not at the same time being distracted by needing to negotiate with other drivers. Whatever the hazard, drivers should be able to understand what it is (Fig. 3.13).

3.13 Drivers need to see and understand a hazard

3. Drivers need to be able to react in time and drive safely

Once drivers are aware of the possibilities of a hazard and can see and understand it, the likelihood of their being able to react safely depends largely on their speed. As noted earlier in the chapter, speeds of 20mph give drivers more chance to be able to react than at 40mph. As an indication, the recorded average speed of vehicles at the complex road junction at Poynton (Chapter 9) is some 12mph. This allows drivers to quickly stop at any position within the junction if necessary. A second indication is the 12mph (20kph) speed limit required by Swiss law at their 'Encounter (pedestrian priority) Zones': a speed that allows drivers to safely respond to and stop for pedestrians in the road (Fig. 3.14). At 10mph a driver can stop within five metres: the length of a typical car (Fig. 3.09).

However, at the basic level drivers instinctively change their driving behaviour when they turn from a spacious, high-speed road to a densely packed urban road. In this sense these are self-explaining roads (Fig. 3.15).

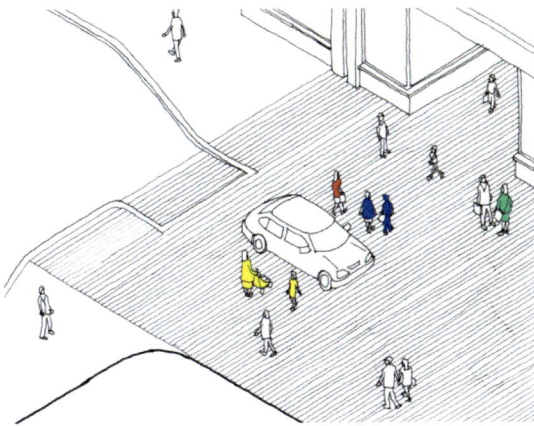

3.14 Drivers need to be at a speed that allows them to be able to react in time and drive safely

3.15 Regardless of traffic signs, drivers tend to adjust their driving behaviour and speed to what they feel is safe, be it a wide, clear road across open country or a congested road in a town

Less-regulated traffic

The principles of less-regulated traffic are explained in Chapter 2. The advantages in streetscape terms of applying the principles are very considerable as there is less reliance on standard traffic management signs (including road-marking lines, diagrams and messages) standard equipment, and street furniture such as Belisha beacons and signal-controlled crossings. The 'mist' of street clutter that obscures or reduces the streetscape quality of many streets, mentioned in Chapter 1, is not required. The case studies in Part 2, as well as other safety studies such as at Broadway, Bexleyheath (Note 3.6), demonstrate that when applied correctly, they can be as safe as conventionally regulated traffic schemes.

There are numerous well-established examples in Europe, particularly in Denmark and Holland. In Switzerland the scheme at Solothurnstrasse, Grenchen, applies new pedestrian priority traffic rules to an existing street while at Bahnhofstrasse, Biel, close by, the road surface of the town square has also been completely rebuilt. In the UK, apart from at Home Zones and Quiet Lanes (Chapter 2), there is no specific traffic legislation to encourage less-regulated traffic, so any experimental schemes are carried out under existing traffic management rules. The case studies in Part II assess what can be done within existing legislation and traffic rules.

Legal responsibilities and liabilities

Drivers tend to modify their speed according to what they perceive is safe, so there are questions as to the necessity of some traffic warning signs. Though they are illustrated and scheduled in the Highway Code, signs that give warnings are not mandatory. However, local highway authorities have duties regarding road safety and although some highway authorities systematically remove what they consider to be redundant warning signs, there has been discussion as to the liability of highway authorities regarding warning signs. The case of Gorringe v Calderdale (Note 3.7) concluded that there was a danger of sign overload and the limited effectiveness of too many signs.

The case concerned a regrettable accident that occurred at a hump-backed bridge when a vehicle driven by the appellant collided with another vehicle coming in the opposite direction.

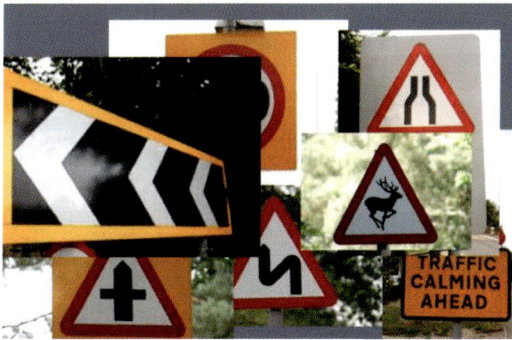

3.16 Warning signs are not mandatory and can be a distracting information overload

3.17 Mandatory signs that give essential information are normally mounted at a high level where they can be easily seen. An arrangement that fits neatly into the streetscape at Coventry city centre is discussed at Chapter 7

The appellant claimed that the local highway authority had been negligent in not reinstating a warning sign. The judgment by the Law Lords was that the road was obviously dangerous and there was no need for a sign. They considered that drivers should take the road as they find it, and be responsible for their own safety and the safety of others. There was a danger of there being too many signs.

The judgments of that case are supported by evidence of driver perception. A sign that ceases to have relevance, such as giving a warning of something that rarely happens, such as a flood, is disregarded. A sign that warns of something that can easily be seen such as a humped back bridge, is also ignored. A clutter of irrelevant non-mandatory signs (Fig. 3.16) reduces the impact of the few signs that are really useful or necessary. Redundant traffic signs are being removed at a number of places across the country. Norfolk County Council has a rolling programme to do so as part of normal highways maintenance, thus reducing future costs. In Surrey, parish councils carry out initial surveys based on a risk assessment and work with the county council highway authority to remove them. Signs that *are* considered essential or are mandatory can be inserted neatly into the streetscape and be seen and understood more easily by drivers (Fig. 3.17) because a clutter of unnecessary signs does not surround them.

SAFETY AUDITS AND QUALITY AUDITS

A matter that is frequently raised concerning the design of traffic schemes is the role of a safety audit that is applied to proposals during the design process. Though normally considered a purely technical process, it can have an adverse effect on a streetscape if it is applied without consideration for the wider objectives of a traffic scheme as outlined in this book.

Official advice (Note 3.8) stresses that a balanced view is needed. A safety audit is not a procedure that passes or fails a scheme. It is intended to form an important part of a wider consideration of all aspects of a design, including streetscape, that are brought together in what is termed a quality audit. Audits should continue after the scheme is in use and include an assessment of the evidence of road safety, so that if necessary adjustments to the scheme can be made.

Evidence-based design

Any discussion on road safety must be underpinned by facts. Often it is said that at a certain location an accident is waiting to happen, but the fact that drivers are aware of potential danger might be sufficient for them to take extra care to avoid an accident – as we have seen, roads that appear to be safer than they really are will have the most frequent accidents. Designs that are based on evidence are the most sound. Road safety data is readily available in order to build up nationwide or local knowledge on road safety. As noted on page 43, the Department for Transport publishes a wealth of information, including an annual review of national data with analysis and details on national trends, causes, variations according to road user, time of day and day of the week. Details of accidents at specific sites are accessible on the Crash Map website (Note 3.9) that records all accidents known to the police by date and by precise location. Information on traffic flows on main roads are available on the traffic-counts website (Note 3.10).

To complement a streetscape analysis it may be useful to check traffic flows and speed on a road, if only to compare the data at a local street in comparison with known streets elsewhere. Although highway authorities and local police will have information on traffic flows and actual traffic speed on most roads in their areas, it may be useful for members of local groups, town and parish councillors – in order to better understand a local traffic system – to know the comparative flows and speeds at adjacent roads and junctions (Box 3.1).

Box 3.1 Understanding basic local traffic data
First it is helpful to understand the role of a particular road in the wider network (Chapter 2, page 28). Then the two key facts:
1. Volume, expressed in the flow of car units every 24 hours and
2. The speed that 85% of vehicles are not exceeding, expressed in miles per hour, (termed the 85th percentile speed).

1. Volume
Most UK roads carry their maximum flows on Friday evenings. This is typically 10% of the 24-hour flow. A quick calculation of multiplying a 10-minute flow observed at 5.00pm by 60 will give an approximate but useful indication for comparison purposes of the 24hr flow.

2. Speed
There are a number of inexpensive radar systems. The Pocket Radar has the advantage of being quite small. Other more elaborate systems include recording devices.
There is a difference between trying to reduce speed with signs and yellow clad people using obvious equipment at the roadside, and simply measuring existing traffic speed for statistical purposes. In these cases it is important that drivers do not notice that their speed is being measured as it may affect how they drive.

Conclusion

In this chapter we have stressed the importance of road safety, have examined how accidents happen and reiterated that there is seldom a single reason, rather accidents occur when several factors happen at the same time. The road itself is seldom the main cause but driver behaviour is usually a contributing factor. By driver behaviour we specifically referred to the ability of a driver to:

1. Expect a potential hazard
2. See and understand it
3. React in time and drive safely.

We noted the severe limitations of the human eye and the difficulties in being able to understand a situation. Bearing in mind the complex tasks of perception, there is a strong likelihood of drivers suffering from inattentional blindness. Ultimately they need to be able to react in time to a situation in order to drive safely. Regarding the ability to react, we saw the significant effect of increased speed on the time taken both to understand a situation and then to physically stop a vehicle.

We looked at the concept of self-explaining roads that help drivers to expect, see, understand and react safely to potential hazards on roads possibly designed in combination with a landscaped traffic calming scheme or where traffic is less regulated, as described in the previous chapter. This approach allows innovative traffic arrangements to complement or even enhance the streetscape character of a place. As a result there is less reliance on traffic-related equipment and street furniture, including road signs. Despite a highway authority's legal responsibilities and liabilities, it is possible to reduce the clutter of conventional warning signs. Finally, evidence of the number and nature of accidents at any road or junction in the country can be easily checked so that comparisons of safety records can be made.

We have seen that streets can be attractive, provide for efficient movement and create conditions where drivers are less likely to make errors of judgment and cause accidents. In the next chapter we look at how modern streets are expected to cope with numerous people-centred activities. The challenge is far greater now than in Gordon Cullen's day.

"

Everyone should be able to use our streets. RNIB's campaign Who Put That There helps people with sight loss to lobby local authorities to reduce transient clutter: advertising boards, pavement parking, bins and rubbish. Whereas a limited number of permanent objects such as benches and lampposts provide friendly landmarks."

Kevin Carey
Chairman, Royal National Institute of Blind People

"Our programme to encourage children to Walk to School brings benefits for health and wellbeing, the environment and road safety. With more pupils walking to school, car congestion around the school gate is reduced. Children who walk enter the classroom feeling alert and ready to learn."

Archie Robertson OBE
Chair, Living Streets

4.01 Living Streets – Walk to School campaign

USER-FRIENDLY DESIGN

CHAPTER 4

Having dealt with visual quality, efficient movement and road safety, we now consider the improved quality of life that a street offers to all sorts of people, including those with disabilities, doing other things as well as just passing through. The need to construct accessible and inclusive streets is seen by government as a golden thread running through all streetscape and public realm schemes. Local authorities have a public sector equality duty under the Equality Act 2010 that includes having due regard to eliminate discrimination and advance equality (Note 4.1).

This chapter examines how a street can accommodate all these additional activities, such as children walking to school (Fig. 4.01) and the equipment they need, and how they interact with each other. At the same time it can be part of a pleasant, enjoyable and distinctive streetscape within a city (Fig 4.02), town, suburb or the countryside.

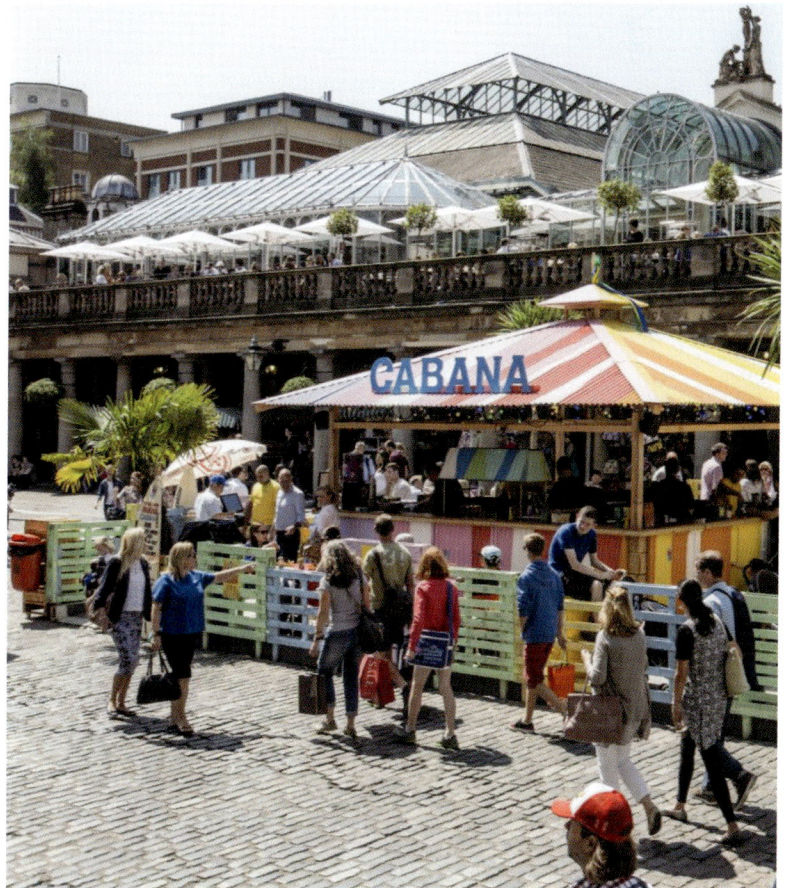

4.02 Temporary stalls and pop-up cafes add delight and interest to a streetscape. Covent Garden, London

Walking

Apart from traffic the most basic use of a street is for walking. It can be enriched by being in a pleasant, interesting place but made uncomfortable, or even impossible, for people with disabilities, if it is obstructed by too many café seats, parking machines and advertising boards, or simply left looking uncared-for when there are no people about (Fig. 4.03). The principles that we examine next are to what extent different activities can take place without precluding others. A seaside sandy beach is an example of a place being used for many activities without the need for permanent equipment. The people themselves bring everything they need so that they can walk, run, sit, have picnics and play games, and at the end of the day they take away their seats and windbreaks, and the tide comes in to clean the whole beach (Fig. 4.04). In a city, given reasonable weather, people will use whatever is available to sit on, even the steps of a cathedral. It doesn't need to be specially constructed formal seating (Fig. 4.05).

4.03 Parking ticket machines and signs insensitively positioned and uncared-for

4.04 People on a beach manage to enjoy themselves without having the clutter of permanent structures. Whitby, Yorkshire

58

In a city street there are other techniques: instead of totally removing equipment when it is not needed, it can be lowered under the ground out of sight, combined with other equipment or neatly fixed to buildings. This chapter considers the practicalities of each activity in turn: walking, cycling, waiting for public transport, and driving and parking cars, also bearing in mind the needs of people with disabilities. Streets are also places where people socialise, meet friends by chance or by arrangement, sit, relax in street cafes, watch street entertainers, visit out-door art exhibitions and street markets. People expect their streets to be attractive, appropriately landscaped with street trees and pocket parks (mini-parks created in small spaces) both temporary and permanent, and want to feel safe and secure in a place that is looked after and well managed. At the same time streets are places where people work; they include delivery drivers, road sweepers, street equipment maintenance crew, and the teams engaged in laying paving slabs and road repairs, and digging up the road to work on underground pipes and cables.

4.05 People sit on whatever is available, such as the steps of St Paul's Cathedral, London

USER-FRIENDLY DESIGN

Campaigns to encourage walking emphasise that people are more likely to choose to walk every day if the route is direct, has interesting views and vistas, and is safe and generally pleasant. The interaction of life and human activity in the street (Fig. 4.06), shop displays and well-kept landscape is far more pleasant than walking past oversized advertisement hoardings and unkempt abandoned sites. Mental health may be improved when people begin to feel part of the scene, see familiar faces regularly and even stop to chat (Note 4.2).

Walking in the countryside is accepted as a leisure pursuit. Walking in towns is a more acquired taste but is made more enjoyable in attractive surroundings, which includes the quality of the pavements. Smooth natural stone slabs, less expensive concrete slabs or a low-cost blacktop finish can all be laid as practical and elegant surfaces. Much depends on adequate maintenance and sufficient budgets to cover the damage to a pavement likely to be suffered at a particular location by vehicles passing over it or service trenches being dug. As mentioned above, uncluttered pavements are essential, and the techniques to achieve them by reducing the need for posts, boxes and other equipment to be put on pavements are dealt with in subsequent sections of this chapter. Examples of pavement surfaces that are practical and considered visually appropriate to the streetscape of cities, towns, suburbs and the countryside are assessed in the case studies in Part II.

4.06 Someone walking to work is likely to see others already at work

Cycling

People who walk or run to keep fit seem to be able to blend into the scene quite seamlessly as they do not need any special signs or equipment that reduces the attractiveness of the street. Most other activities do seem to generate street clutter, unless care is taken to reduce it, and cycling is an important example of this. There are strong reasons to promote cycling: it provides relatively stress-free exercise, is a good form of transport for local journeys, and allows people to breath fresh air and enjoy the scenery. However, as cycling becomes more popular the demand for special provision increases, as does street clutter such as parking and signs. The challenge is therefore to help cyclists use streets without the need for more overwhelming infrastructure, equipment and signs.

Even when cycle lanes are provided, many may find it more direct to use an adjacent road. Others may feel they need as much protection as possible and value dedicated cycle lanes. However, there is an obvious practical limit to these and there will be places where cyclists need to use ordinary streets where the traffic speeds are 20mph or lower, possibly with built-in direct routes, including riding against the normal flow of traffic. It is usually quite obvious where cycles are intended to be, so dedicated signs can be used sparingly (Note 4.3).

CYCLE PARKING

Cycle parking is a great problem for most cyclists, as there is usually not enough available space and cycle parking adds significantly to street clutter. Cyclists naturally want to park their bikes at the end of their journeys. In extreme cases such as in Amsterdam some cycle parks almost have the appearance of untidy heaps of tangled steel. Where the numbers of parked cycles are modest they can be accommodated in the road on space otherwise used for car parking (Fig. 4.07), rather than on the public pavement where people walk.

4.07 Cycle parking neatly positioned on the road well away from where people expect to walk

Public transport

Public transport on streets usually means buses, though it includes taxis, trams, coaches and even buses that are designed to look like trams. In this chapter we do not have the space to consider all the complex details of public transport management but will restrict the discussion to how the various forms of transport can be integrated into an attractive place.

BUSES AND TRAMS

Possibly the most common piece of public transport equipment is the humble bus shelter. Many styles are available – they do not necessarily need to be of a quaint pseudo-historic style to fit into a historic street scene. The essential design objective is to bring together as much as possible of the other equipment that is normally needed at a shelter. Seats, bins, signs, lighting, timetables and ticket machines can all be assembled in a single comprehensive whole (Fig. 4.08). In some places shelters can be made part of an adjacent building.

Streets are continually being retrofitted with new equipment to accommodate the most recent innovations in transportation. In order to reduce the likelihood of ever increasing street clutter and visual chaos, a standard product that might have been used at one location may need to be adapted before it is acceptable or fits in with other design requirements at another. For example, the design of bus lanes and bus gates can be unobtrusive so that they fit neatly into the street scenes. It is also possible to install a tram system without the need for overhead wires, cables and support posts.

4.08 Bus shelter with timetables, seats and a green roof. Greater Manchester

Personal motor transport

Personal motor transport usually means cars and because for much of the time they are stationary, car parking is an important streetscape issue.

In some places, street space is so scarce that car parking is restricted to off-street car parks. But on-street parking is common, and drivers need to know where and when they should park. There are more attractive alternatives to the normal ugly yellow lines. Restricted Parking Zones have simple signs at the zone entrances, and parking places are indicated possibly by a distinctive road surface (Fig. 4.09), rather than by yellow and white lines. A large zone in Coventry is described in Chapter 7.

Where the times for parking or deliveries need to be explained, the signs can be fixed neatly to existing private railings and walls (Fig. 4.10). London traffic authorities have the power to do so (Note 4.4). Elsewhere it can only be done with the agreement of the property owner.

Depending on the need for space, people tend to park their cars wherever they think they are safe. At Poynton (Chapter 9) the road is only one lane wide in each direction and there are parking bays, though no road markings. Drivers tend not to park on a single lane because it blocks the whole road. At Poundbury (Chapter 10) the road varies in width to allow people to park conveniently near their houses, and no special rules or road markings are needed.

Parking bays are handy lengths of road that can be used for different purposes at different times. One length of road can be used as a delivery bay in the morning, and a temporary café in the afternoon and evening. Under-used parking space can be turned into a cycle-parking area, temporary green space or pocket park. Delivery bays can be on the pavement and form part of it when not being used for deliveries.

ELECTRIC CARS

As technology and social concerns change there are constant demands for new equipment to be put in streets. This simply adds to street clutter. But there are many opportunities for innovation in the design of street equipment, and each solution should respond to the local needs and character of the locality. This is currently the case with regard to helping the drivers of electric cars by providing power supply pillars on the pavement at parking places (Fig. 4.11). In time a version of the pillar may be available that can be lowered below ground level when not needed, as is currently possible for electrical supplies to temporary market stalls.

4.09 Distinctive road surface indicates parking places. Bury St Edmunds, Suffolk

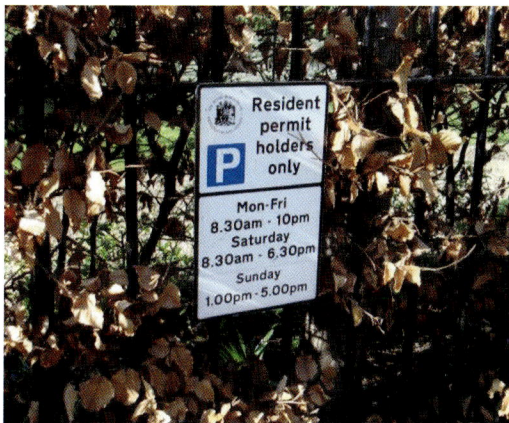

4.10 Where parking signs are needed, they can be fixed neatly to walls and railings. Kensington, London

4.11 Hopefully future arrangements may be tidier

63

People with disabilities and vulnerable road users

A frequently mentioned streetscape issue is the installation, at what seems to be almost every street corner in the country, of lines of blisters about two centimetres wide set into the pavement, sometimes on concrete or stone paving slabs, coloured bricks or as stainless-steel domes. Known as tactile surfaces, they are to warn people with visual impairment of the possibility of traffic and also to help them find a safe place to cross. They are installed at crossing places where the pavement is ramped down to road level to help people who have difficulty walking or who use wheelchairs. Unfortunately the ramps remove the kerb that people with visual impairment rely on to indicate that they are still walking on the comparatively safe, raised pavement. And the blisters are an irritation to people who have walking difficulties or use wheelchairs. In fact at many street corners the configuration of tactile surfaces appears to be quite incomprehensible (Fig. 4.12). This dilemma has been known for several decades but there seems to be no simple remedy.

However, there are neater ways to install the paving and keep to the national guidelines (Fig. 4.13). At a street corner that includes a tight kerb radius, which is easier for people with disabilities to use, the tactile surface is restricted to the ramped paving, where it also acts as the guide to where it is safe to cross (Figs. 2.25 & 2.26).

National consistency is important but sometimes, to fit local circumstances, national standards can be adapted with the knowledge and agreement of local disability groups. Street clutter that obstructs the pavements should be reduced to the minimum. But as Kevin Carey, Chair of the RNIB states, a limited number of permanent objects such as benches and lampposts provide friendly landmarks. There will inevitably be some places where the assistance of a sighted person may be helpful.

Though not strictly speaking people with a disability, children are vulnerable road users and so many are prevented from going out on to a busy street on their own until they have reached secondary school age. To be able to cross a road safely is as challenging for children as it is for adults with disabilities because though children may be able see quite well they are less able to understand the speed of a vehicle or its likely movement. Most streetscape designs that help people with disabilities will help children too.

4.12 Remarkably ugly and difficult-to-understand tactile surfaces demonstrate the complexities of applying current guidelines

4.13 Tidily installed tactile paving and the neater narrow version of yellow lines

Recreation

People like to relax. Sitting at pavement cafes, seldom seen in the Britain of Cullen's day, is now common across the country. Private cafes may now occupy public pavements, previously intended for walking, and whereas this may be justified, there are very many demands on the use of pavements for purposes other than walking, and on roads for uses other than traffic. The demand on space is so great that some form of land allocation or multi-use becomes desirable. The principle is to make sure that each activity does not need any permanent structure or equipment that will remain when not needed, inconveniencing or preventing other activities and making the place less attractive. As an alternative to permanent equipment, waste bins, urinals, floodlighting, fountains and performance areas can be set level with the ground or lowered out of sight when not in use. A contrast in the use of a space can itself be attractive. For example a market square might be the scene of lively noisy activity on a Saturday night, and after clean and tidy-up the same space can become the quiet backdrop to a relaxing stroll the following morning.

STREET MARKETS AND ENTERTAINMENT

The street has a long tradition as a place of occasional entertainment and casual trade, such as an ox roast. Today these traditions survive or are revived, through using more hygienic methods. Current health-and-safety requirements as well as expectations of modern convenience and comfort mean that even a modest temporary market stall may need electricity and water. Permanent structures such as a raised dais tend to make an area suitable only for the limited use intended for the structure. Temporary stands and stalls may be more practical, and some markets are made up entirely of vehicles that can easily be driven away when the market closes (Fig. 4.14). Traditional market stalls can be given water and electricity from stands that lower into the ground when not needed. Layout arrangements can be marked on the ground by neat metal studs rather than painted white lines. The tradition of street performers is more easily accommodated in a modern street, but one person's entertainer is another person's nuisance, so the local authority may need to have a licensing system and make arrangements for special cleaning.

4.14 Mobile vendor van adds life to a street and can be completely removed when not needed

Street trees and landscape

"Street trees can improve your mental health" (Note 4.5). Opinions and studies stretching back some hundred years stridently proclaim or tentatively suggest that trees and urban greenery reduce stress. Victorian philanthropists undertook programmes of street tree planting for the public good. Twenty-first-century studies find a causal link between a reduction in anti-depressant prescriptions and the presence of street and public space trees. Each spring at cherry blossom time, Japanese families take part in the festival of Hamani and picnic under the trees. Certainly in the UK people with choice appear to live where there is greenery.

A basic layout design is a line of street trees. Even within this simple theme designs can vary so that the species and position of the trees relate visually to adjacent buildings. An example is a line of silver birch trees such as those seen in Kensington (Fig. 4.15). Their white barks work well with white painted stucco. Trees are often planted along a street almost at random. But with some thought they can be positioned to be part of a co-ordinated design related to adjacent buildings, as green walls or on the buildings themselves (Fig. 4.16). However, there are also other layouts to consider. In the terraces of Georgian Bath and Edinburgh New Town, trees were consciously not planted along the streets. They were reserved for the green squares and gardens, planted where they had the space to mature in what are now groups of towering majestic specimens forming the splendid focal points of grand vistas.

4.15 The white trunks of the silver birch trees echo the white columns

TREE CARE AND MAINTENANCE

As living organisms, trees keep growing, both under the ground as well as above. Large forest trees, for example, need to be planted with care because of their potential size. The possible damage to the foundations of buildings and underground pipes and cables causes understandable concern. Tree pits for street trees can be lined to prevent tree roots spreading to damage adjacent property and underground equipment. A street tree needs care during its life: as it grows stakes need to be removed and iron grills adjusted. Local procedures are needed to make sure that every tree has someone who can arrange for its care and maintenance.

4.16 Co-ordinated greening of roads and buildings. Singapore

Safety, comfort and maintenance

Safety and security is a fundamental requirement in a civilised street. A place needs to *feel* safe as well as *be* safe. Fear for personal safety deters many people from venturing out alone at night. A place that is clean, tidy, well-lit at night, seems to be looked after and has obvious signs of human life will feel safer than a place that is dirty, full of rubbish, has graffiti, is poorly lit, has abandoned or untidy excavations in the pavement, and feels generally neglected.

GRAFFITI AND RUBBISH

The local authority can deal with graffiti and rubbish removal as well as street sweeping. There are national standards for the removal of street litter according to the use of a street (Note 4.6). Graffiti are a telltale sign that no-one cares about a place, so need to be removed promptly. Similarly unsightly commercial rubbish left on pavements to be later collected is unnecessary. Procedures, sometimes put in place by Business Improvement Districts or groups of businesses, arrange for rubbish to be collected locally in a building, from where disposal contractors collect it.

SECURITY CAMERAS

The presence of security cameras, rather than giving a sense of security, might make someone feel that the place is unsafe and therefore needs cameras. If they are required, security cameras need not be too obvious. Rather than being fixed to stand-alone posts positioned in the middle of a pavement, they can be discreetly positioned on existing buildings.

STREET LIGHTING

Street lighting is not only for the needs of drivers; it helps with security. The cluttered effect of too many lamp columns can be reduced by fixing lamps to existing buildings (Fig. 4.17). In the City of London this is a requirement considered at the building planning stage. Other London local authorities have similar recent powers to place street lights on private buildings (Note 4.4). Elsewhere, local authorities need the agreement of the property owner. Innovative lighting schemes can retain historic street gas lamps by supplementing the limited illumination of the gas lamp with lighting from shop front displays and unobtrusive auxiliary lights at doors and features facing the street.

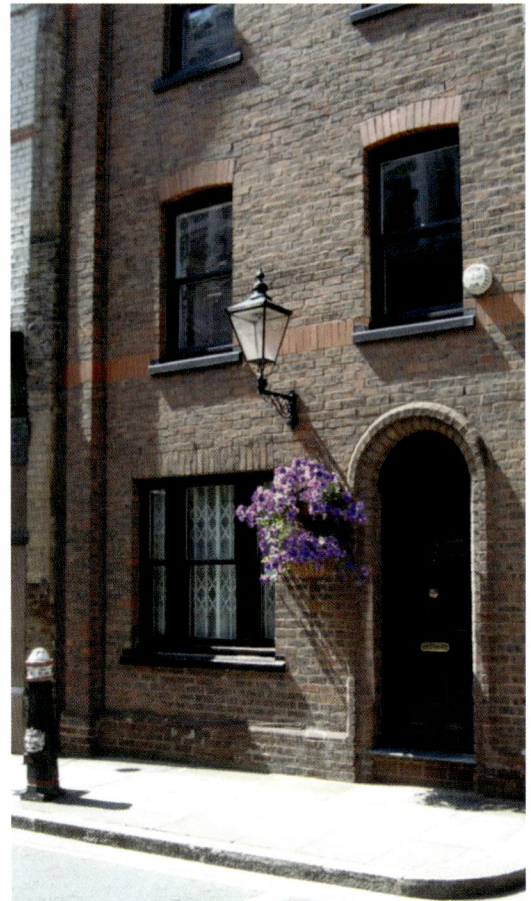

4.17 Public street lighting fixed neatly to buildings. Middle Street, City of London

ON-GOING CONSTRUCTION AND MAINTENANCE

All streets need constant maintenance – holes in the road – ranging from major excavations and traffic holdups to the constant minor repairs of underground service pipes and cables. The streetscape consideration depends on location. In the countryside and suburbia the work would be sufficiently infrequent for no special visual consideration to be necessary. But in dense urban areas such as a city or town centre special screened barriers to protect workers and fit acceptably into the streetscape are helpful. These can be temporary green walls, decorative screens or in the case of building construction site, a tromp-l'oeil illustrating at full size the façade of the proposed building (Fig. 4.18).

4.18 Building construction works concealed behind a temporary trompe-l'oeil

NEGLECT

The most common signs of neglect in a street scene are an abandoned piece of equipment or redundant sign. It is surprising how a lick of paint or removal of a redundant sign can improve a place. Co-ordination is difficult because so many agencies are involved, and few will feel able or willing to touch the signs or equipment of others. However, as we explain in the final chapter, other groups such as Business Improvement Districts or local voluntary organisations may be able to help, if not by carrying out the work themselves then by drawing a problem to the attention of the relevant body. Often making sure that something is painted and is being looked after is more important to a streetscape than deciding its colour.

Co-ordination

A key aspect in improving the quality of a streetscape is some form of co-ordination. This is provided to a degree by the town and country planning process but planning control does not cover most of the matters discussed in this chapter. The sheer number of activities, interests and separate priorities that a modern street has to accommodate can be overwhelming. It seems that every day a new group is formed to promote a particular use of scarce street space; there are separate expert professional practitioners engaged in the provision of the different activities seen in a street, and the funding and management of a street is quite fragmented. This is unlikely to change in the foreseeable future. As we explain in the next chapter where co-ordination exists it is carried out voluntarily by joint council committees, through business groups or in some places by landowners.

Conclusion

This chapter has explained the wide range of design objectives, in addition to being attractive places and carrying vehicles efficiently and safely, that modern streets are expected to fulfil. It examined how those objectives can be achieved while at the same time respecting or enhancing the distinct character of an individual streetscape.

The most basic requirement would be that people, especially those with disabilities, who use a public pavement – the most common activity – can do so in comfort and safety unobstructed by the clutter of service boxes, posts, equipment or just rubbish. Ordinary pavements are where people meet their friends, often by chance, appreciate views and vistas, and find interest in shop displays and the human activities taking place around them. To this end cycle parking can be neatly positioned on the road at the kerb rather than on pavements, on-street car parking can be tidily located without the need for yellow lines, possibly among trees and landscape, and bus shelters can be provided with incorporated seats, real-time information boards and bins. There are opportunities for shelters and other essential street furniture to be designed, as was suggested for pavements in Chapter 1, to complement the streetscape categories of city, town, suburb or the countryside. Whatever the detailed design of a streetscape, it should be carried out with needs of people with disabilities in mind, and particularly to help people who are blind or have visual impairment to cross a road.

Streets are used for recreation and people expect there to be street cafés, food stalls and landscape, many of which can be fitted into existing streets at a flexibly-used kerbside where they might be temporary pop-up spaces, and at other times used for car parking. But the underlying essential ingredients of successful streetscapes are basic security, cleanliness, tidiness and a feeling that someone cares. Security cameras can be unobtrusively mounted on existing buildings rather than on separate posts, rubbish need not be stacked up on pavements, and street lighting can be efficient and designed to complement each of the streetscape categories. Examples of these practices are examined in the case studies in Part II and the practical systems that finance that deliver all the ingredients of a successful street mentioned above are considered in the next chapter.

"

From small beginnings some twenty years ago the Business Improvement District concept has taken hold. Through it local businesses engage and contribute directly to the promotion, security and vitality of their locality. The number of BIDs is growing: evidence that businesses are convinced of their financial value."

Ruth Duston
Chief Executive,
Northbank Business
Improvement District

"Covent Garden was designed to bring people together. We continue that legacy with innovative art installations, events and an exciting combination of shops, restaurants and amenities. By maintaining our streetscapes with painstaking attention to detail, we create one of the world's best places to live, work and visit."

Andrew Hicks
Director of Estates,
Covent Garden,
Capital & Counties Properties PLC

5.01 A dramatic public art installation paid for by the private sector at Covent Garden, London

FUNDING AND COMMERCIAL VIABILITY

CHAPTER 5

The previous chapters have established that it is possible to create attractive streets that respond to their context, provide efficient and safe movement of traffic, and accommodate the increasing demands of a modern society. This chapter deals with how streets are funded and how the quality of a street can affect the viability of businesses in a locality.

5.02 Heritage street lights at Covent Garden supplemented by shop-window lighting

Who pays?

Each year more than £1,500m (Note 5.1) is spent on maintaining and improving the road network. The projects range from filling in the smallest pothole to a substantial multi-million-pound scheme, and are funded by the public sector, by the private sector (Fig 5.01) or jointly, by both (Fig 5.02). Funding for new or improved streets normally comes from public funds gathered through general taxation. These funds are allocated from central government to local highway authorities, and other councils and agencies. Though funding is remarkably complex in structure, there seems to be no trend towards simplification, so anyone interested in influencing decision making regarding public funding needs to be aware of the systems in place at a particular location. Some guidance on this is given later in the chapter.

In parallel to public funding, often as supplementary or partial funding, are systems to contribute shared or individual private finance. These funds may be paid by businesses as part of planning gain agreements such as Section 106 agreements and the Community Infrastructure Levy (CIL) associated with new building developments. They may also be provided by individual local organisations and companies, or through umbrella organisations such as Business Improvement Districts (BIDs). Local businesses in a BID may agree collectively to contribute money towards street improvement projects that they feel will be of commercial benefit. These groups of businesses as well as the landowners of some large urban estates act in a similar way to the eighteenth and nineteenth century estate developers mentioned in Chapter 1. They appreciate the long-term as well as the short-term commercial advantages of good street design.

FUNDING THROUGH THE PUBLIC SECTOR

Though each local authority collects rates from householders and businesses, with few exceptions these funds are transferred to central government, combined with general taxation and allocated to each local authority, under budget headings, according to local needs. These government funds are distributed through a number of departments to local councils and other agencies (Fig. 5.03). The majority is spent on statutory services such as highway maintenance, local highway improvement, street lighting and road safety that the authorities are obliged to provide. Some locally collected funds such as for car parking may be spent locally. As budgets are allocated for specific purposes, they tend to be self-contained and focused on a single issue, such as road safety or street lighting, and apply at different times-scales, so that there is seldom any formal long-term coordination. This has a profound effect on the appearance of a street.

However, there are exceptions. Very large projects usually have multi-disciplinary teams of experts who coordinate them. This combines street lighting, traffic signals, parking, traffic management and landscape, etc. to achieve very satisfactory results. In some places voluntary agreements bring together the relevant agencies. Though as time elapses after completion of a scheme, unless extraordinary measures are taken, the normal ad hoc arrangements for continual adjustments gradually erode the qualities of the scheme.

DELIVERY SYSTEMS AFFECTING STREETSCAPES

Government ministries & departments for:

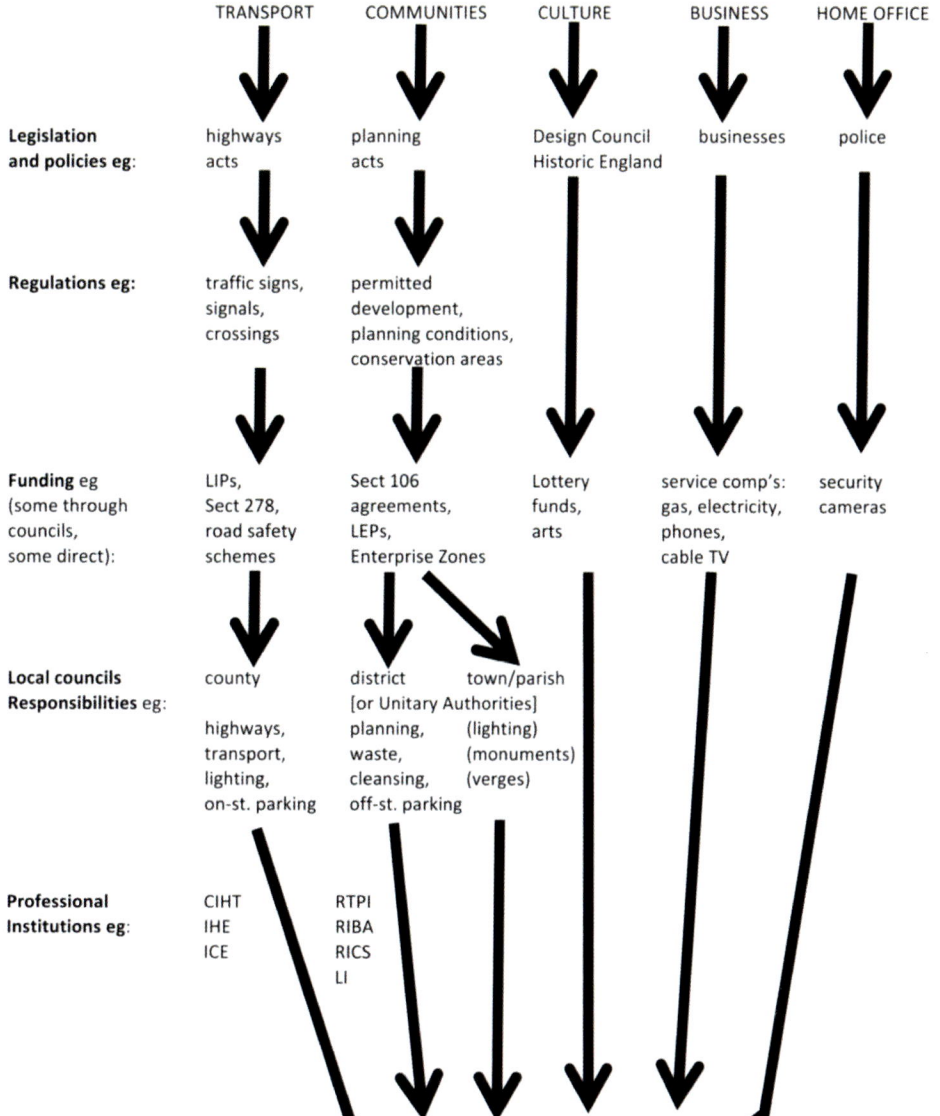

TRANSPORT COMMUNITIES CULTURE BUSINESS HOME OFFICE

Legislation and policies eg:

highways acts | planning acts | Design Council Historic England | businesses | police

Regulations eg:

traffic signs, signals, crossings | permitted development, planning conditions, conservation areas

Funding eg (some through councils, some direct):

LIPs, Sect 278, road safety schemes | Sect 106 agreements, LEPs, Enterprise Zones | Lottery funds, arts | service comp's: gas, electricity, phones, cable TV | security cameras

Local councils Responsibilities eg:

county

highways, transport, lighting, on-st. parking | district [or Unitary Authorities] planning, waste, cleansing, off-st. parking | town/parish (lighting) (monuments) (verges)

Professional Institutions eg:

CIHT IHE ICE | RTPI RIBA RICS LI

Key to government ministries and departments:
MHCLG: Housing, Communities and Local Govt.
DCMS: Culture, Media & Sport
BEIS: Business, Energy & Industrial Strategy

Key to funding sources:
LIPs Local (transport) Implementation Plans
LEPs Local Enterprise Partnerships
Sect 106 (town planning) agreements
Sect 278 (highways) agreements

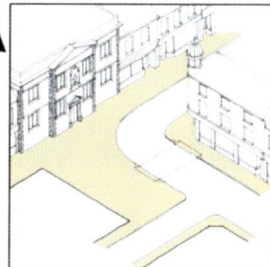

A typical area of the public realm is affected by at four or five government departments, each with its own policies, powers and funding streams.

Even modest interventions need to be carried out in the knowledge of this context.

5.03 Government funds are distributed through a number of departments to local councils and other agencies. A fuller list of professional institutions is on page 140

WHO DECIDES HOW PUBLIC FUNDS ARE SPENT?

Though at least five government departments are involved (Box 5.1), the far greater allocation of public spending on streetscape matters is from highways budgets, and it is strictly controlled by central government. Government ministers establish priorities and national strategies, such as reducing congestion, and improving road safety, passenger transport and the safety of cyclists. Members of local councils that are highway authorities, such as county councils and unitary authorities, prepare their own programmes known as Transport Plans, or for London boroughs Local Implementation Plans, based on the national strategies, and apply for central funds in order to carry them out.

Government ministers may also establish specific funds, for example for sustainable transport that local highway authorities can bid for. Local councils may also access other specifically targeted funds available through Local Enterprise Partnerships (centrally funded establishments set up to promote business in a specific location) as well as planning gain funding and infrastructure levies associated with new development.

It can be seen that the arrangements are quite complex. They vary according to location and the details constantly change as government ministers adjust their priorities and the methods by which they are implemented. So to determine what system is currently in place at a particular location it is necessary to establish if the highway authority is a county council, a unitary authority or a London borough, and if a two-tier system applies. Similarly it is important to be clear about titles and responsibilities. An elected mayor of a unitary authority has considerable power and should not be confused with a mayor of a district council whose role is mostly ceremonial.

Box 5.1 Government departments' influence on streetscape

Department for Transport (DfT)
The main source of funding for streets is from the DfT through its settlements to local highway authorities for public transport, new highways, highway maintenance, including pavements and street lighting, road safety schemes and on-street parking.

Ministry of Housing, Communities & Local Government (MHCLG)
The second source of funding is from the MHCLG to unitary and district councils for street cleansing and sweeping, street markets, arts and leisure.

Department for Culture, Media & Sport (DCMS)
Heritage Lottery Fund for urban conservation projects.

Department for Business, Energy & Industrial Strategy (BEIS)
Capital and maintenance projects by service (utility) companies on and under highways.

Home Office
Programmes to improve security and prevent crime through the operation of on-street security cameras, etc.

DETERMINING WHO IS RESPONSIBLE FOR A STREET

Though the arrangements appear to be needlessly complex, they are unlikely to change, so the best approach to determine the system in operation on any particular street is to access the government website 'Report a pothole' (Note 5.2). This refers every householder to their local highway authority website. Here there is detailed information about local budgets, current and proposed highway projects and names, with contact details, of local decision-making committees and elected councillors.

Once a budget has been agreed by the elected councillors of a highway authority, the subsequent day-to-day management of a road maintenance budget and the allocation of expenditure to specific projects that may have a significant influence on the attractiveness of a street might be totally delegated to a local highways manager. There are many variations in organisation between authorities. A highways manager might be a direct employee of the authority, or be within an independent outsourced design and contracting company. Management systems frequently change. Dedicated teams are formed and dispersed. Individual staff members join and leave, and outsourced service companies are replaced. It therefore requires some determination to establish exactly who is responsible for a particular street. But as they are administering public funds, local councillors should be able to help.

In addition, though strictly speaking not using public funds, the service or utility companies provide water, gas, electricity, sewerage, telephone and communication cables under the highway etc. and are generally able to carry out maintenance work to their equipment as they wish. Finally, in many areas there are also parish and town councils or neighbourhood committees that are responsible for minor but important amenity and maintenance projects, such as landscaped gardens and open spaces, street trees and benches. The elected councillors or committee members meeting in public usually make decisions on these matters.

JOINT FUNDING

It is very rare that a designer has a clean sheet to work on. Most street improvement projects involve a small step in a journey of continual improvement. As the funding chart (Fig. 5.03) demonstrates, there are many sources of public funds for work in the public realm. The key is to achieve greater coordination so that each adds value to the others in a way that improves the whole, rather than just a part of the streetscape.

77

FUNDING THROUGH A SECTION 106 PLANNING OBLIGATION AGREEMENT AND COMMUNITY INFRASTUCTURE LEVY

As the public sector is being reduced in size, some aspects of funding and roles of co-ordination are being taken on by the private sector. An organisation that wishes to carry out a large building development is obliged to enter into a legal agreement to pay planning gain funds when planning permission is granted. This is known as a Section 106 (of the Town and Country Planning Act 1990) planning obligation agreement. The funds usually relate to a financial contribution towards infrastructure or services associated with or needed by the development. A study published by the Department for Communities and Local Government in 2014 estimated that in the year 2011–2012 the total value of planning obligations agreed was £3.7bn (Note 5.3). A typical example of Section 106 funding is the improvement works at Poynton (Chapter 9), substantially funded through planning gain payments agreed as part of a planning permission for a supermarket. In some instances a highway authority might use part of its maintenance budget by bringing forward anticipated highways maintenance work to coincide with the main project.

The more recent Community Infrastructure Levy (CIL) is agreed to in addition to Section 106 payments. The levy is allocated to help local projects not necessarily directly associated with the development, such as highway improvements, incidental green spaces and street trees.

5.04 Christmas lights along the Strand, London, paid for by the Northbank BID

5.05 The BID's Ambassadors welcome visitors

5.06 The BID's auxiliary street sweeping team

FUNDING THROUGH A BUSINESSS IMPROVEMENT DISTRICT (BID)

Based on a Canadian concept the, first Business Improvement Districts in the UK were established in 2005. The arrangement, whereby local businesses, though not landowners or freeholders, elect to contribute to public works, is supported by legislation that sets out the system for balloting local companies and to ensure that funds are correctly administered. If a majority of eligible businesses vote to take part, a compulsory levy (usually 2% of rateable value) towards funding projects is paid by all businesses in the area. With this degree of autonomy, support given by businesses to what is an additional tax, though dedicated to promote local business, is an indication of their confidence in the financial return on locally targeted investment. The concept is widely understood in the UK. There are now some two hundred BIDs and they are continuing to spread to small, non-metropolitan areas. Early established BIDs have been renewed and are expanding. It appears that the concept is established as a significant method to raise funds for street works.

BIDs contribute a total of £150m per annum, made up of between £50k to £3.5m from individual BIDs. BIDs are usually run by a small team of permanent staff who buy-in auxiliary professional services as needed and have become adept in raising additional funds from third parties. They are highly skilled in co-ordinating funds and implementing projects through several agencies. The projects that BIDs fund totally would include Christmas lights (Fig 5.04) and trained Ambassadors to welcome visitors, reassure them that the streets are safe and point out places of interest (Fig. 5.05). BIDs also supplement public-sector services such as street sweeping and cleansing (Fig. 5.06). This can emphasise the identity of an area. Some BIDs dovetail their sweeping timetables with the normal services of the local authority. Some simply agree to take total responsibility for selected streets. Projects depend on the size and location of each BID, ranging from minor on-going street maintenance to wholly sponsored capital initiatives (Box 5.2). Understandably many businesses prefer expenditure, that they have contributed to through a BID, should be near their own premises or have a direct commercial advantage. Significantly, in order to give it a competitive edge, there is an appetite to fund projects that improve the wider business area.

Commercial viability:
The Covent Garden story

Whereas Business Improvement Districts levy funds from the local businesses, excluding property freeholders, improvements to the streetscape provide obvious financial advantages for property owners who have a long-term view. These advantages are difficult to quantify, but several of the large historic owners of urban estates such as the Crown Estate and the Duchy of Cornwall Estate at Poundbury (Chapter 10) continue to take considerable care in enhancing their streetscape for long-term benefit. A case worthy of examination is the area at Covent Garden, London (Fig. 5.07), including the half-dozen streets around the Piazza, originally laid out as the first privately financed public square in the UK by Inigo Jones for the Duke of Bedford's new estate development in the 1630s (Fig. 1.10). In the intervening years the ownership of the estate was fragmented and changed hands, and the centre of the Piazza became London's wholesale fruit and vegetable market. The historic architectural quality was officially recognised in the 1980s, and with the market gone plans were put in place to regenerate the area and conserve its heritage. In 2006 new owners acquired the core buildings and have developed comprehensive streetscape design and management techniques.

STREETSCAPE QUALITY ENSURED BY TERMS OF LEASES

The reason for taking this as a case study in the practical application of streetscape improvement is that it is one of the few places where all the ideas mentioned in this book are being carried at the same time. Using their skills and experience in the development and management of purposely built shopping centres, the new owners announced their vision to improve the streetscape in the wider area and set about increasing their freehold interests in order to do so. The result is that by implementing the influence they have as freeholders through the terms of the leases on the exact use and appearance of their tenants' buildings, the company ensures considerable co-ordination. This dovetails arrangements for joint funding and management with the local authority to raise standards in security and co-ordination of traffic management, including on-street parking and access for service vehicles (Fig 5.08), street and amenity lighting.

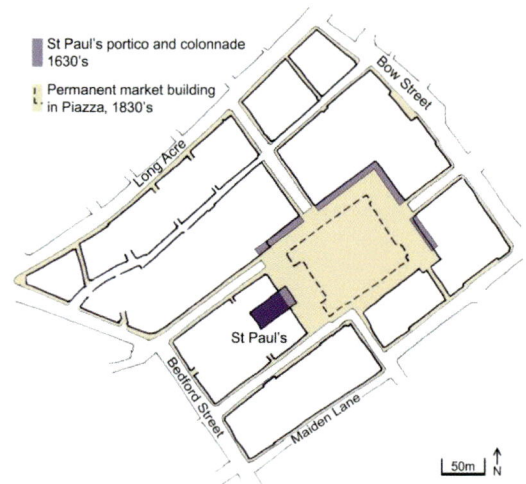

5.07 The Covent Garden area of London is centred on the streets surrounding St Paul's church and the 1630s Piazza with its colonnade (Fig 1.10). The market buildings (dashed line) were added in 1830

Financial involvement extends to pavement quality and on-going street maintenance (Fig 5.09), landscape, informal public seating, tables, umbrellas, street cleansing, litter picking, and commercial waste collection and disposal. The security team has a round the clock presence on the streets and is supported by discretely positioned CCTV cameras. To complete the package there is assistance in the management of world-class street art (Fig 5.01) and entertainment so that the quality of the total streetscape is of a standard seldom experienced anywhere in the UK.

The freeholder's proactivity includes the allocation of key sites as destination attractions and specific streets for specialist retail offers, all with a keen eye for streetscape quality. Shop front design, visually appropriate private and public signage and the reduction of street clutter are handled as a matter of course by liaison and technical advice to tenants. At the larger scale, new public spaces are created within refurbishments or redevelopment, footfall is distributed evenly through new pedestrian routes and major building work is masked while underway. Thus the concept of a sequence of informal interesting spaces, such as at Grainger Town, Newcastle, discussed in Chapter 1, is achieved (Fig. 1.13).

5.08 Trained courteous security and traffic management staff micromanage the public space and effortlessly deal with the public

5.09 Street maintenance work is constant but is carried out behind temporary screens with little inconvenience to visitors

81

STREETSCAPE AS PART OF THE COMMERCIAL OFFER

The area is run and managed to the same quality as would apply to a purposely-built international retail or volume visitor attraction. The difference is that at Covent Garden these techniques apply to real historic streets that are public highways, real historic buildings in an area of documented history where people live and work. All the details of appropriate streetscape designs, explained in the preceding chapters of this book, have been willingly and enthusiastically applied. The endeavour demonstrates what can be done through freehold interests but importantly underlines the long-term economic benefit of quality streetscapes

COMMERCIAL MOMENTUM LEADS TO INNOVATION

The wide range of streetscape enhancement projects concentrated in one location has created a momentum that encourages the company to go further and examine innovations not often seen. The presence of trained, courteous personnel to carry out public security and micromanage service traffic allows informal public seating and tables to be provided. Street lighting maintains traditional London gas lights (Fig. 5.02) but its admittedly low illumination level is supplemented by additional sources of light from shop windows, sensitive lighting at doorways and at the existing traditional London front area railings.

    As in most successful areas, construction is constantly taking place. Shops are being renovated, and groups of buildings are being adapted and extended. Yet although the area has to continue to be attractive to visitors every day of the year and virtually twenty-four hours a day, construction and maintenance needs to be carried out. The larger building sites are surrounded by a tromp l'oeil applied to the whole façade of a building (Fig 4.18). Smaller work sites on the public areas and streets are hidden from view by temporary green screens or, where appropriate, more elaborately designed decorative hoardings. Informal public seating and tables, removed at night, are designed to complement their locations such as shopfronts (Figs. 5.10 & 5.11). At the same time lively entertainers appear (Fig. 5.12) and spectacular artistic installations are exhibited as well as pop-up restaurants and yoga classes for local residents. The neighbouring Royal Opera House's outreach programme takes a taste of international opera and ballet to a wider audience by putting on free public shows in the Piazza.

5.10 Informal seating for the public outside shops

APPLICATION ELSEWHERE

Obvious discussion points revolve around how such a project could be applied elsewhere. The power of freeholders is clearly significant, but grouping freeholders together is challenging because people and companies own properties for different reasons and may not all have a long-term view in mind. There is always the possibility that without a legal entity binding them together, a single freeholder may simply opt out of taking part but still benefit from the efforts of neighbours. Perhaps the main lesson is to encourage local groups, possibly through BIDs, to widen their sphere of influence and raise their aspirations.

5.11 Seats and tables for visitors are taken in at night

5.12 Entertainers at St Paul's church portico are strictly controlled

83

FUNDING AND COMMERCIAL VIABILITY

Economic benefits of attractive streets

Putting actual numbers to the benefits of improved streets is difficult, and there are both enthusiasts and sceptics. The challenge is to isolate a particular improvement and attribute a monetary value to it. Often the public realm is just one factor of an experience. People will put up with poor conditions if there are other reasons to visit an area, such as a range of shops, leisure or sports attractions. Yet experience from the retail sector suggests that people associate a standard of quality environment when they purchase with the nature of the article purchased. The willingness shown by businesses to voluntarily pay through a BID or as freeholders for improvements that would otherwise have been paid for by general taxation, to make a place more attractive, appears to support the proposition that more attractive places mea more successful businesses.

People understand a street as part of a total experience. The activities that take place in the street as well as the design of the street itself and the adjacent buildings are equally important. Where several changes take place at the same time the improvements can be very apparent, though they could of course take place incrementally.

5.13 Coordinated changes and improvements: before and after the removal of on-street car parking and renovation of shop fronts

Shop fronts can be sensitively improved, traffic and parking can be reorganised, and people given opportunities to stop and enjoy themselves, and these all contribute to an improved quality of public realm (Fig. 5.13). At another example, at Stratford upon Avon, the emerging Neighbourhood Plan being undertaken largely by local people includes thoughts that traffic arrangements could be adjusted to transform an ordinary road junction into a place people would want to visit and return to – a sound economic advantage for the local businesses (Fig. 5.14).

The feeling that someone cares is certainly evident when a new scheme or project is first completed. As time passes, if there is inadequate maintenance or cleaning the standards quickly deteriorate, the original expenditure or investment is questioned. For the maximum economic benefit in the public realm, co-ordination of effort and continual appraisal, management and maintenance within a core strategy are essential.

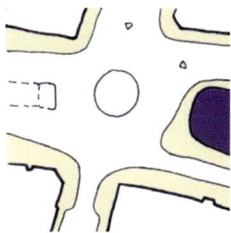

Current
Narrow, congested pavements and conventional roundabout

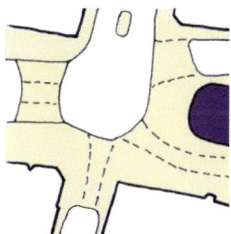

Suggested (also right)
More useable pedestrian space with uncontrolled junction

5.14 There are sound economic benefits when places are created that people want to visit

Co-ordinated professional expertise

A theme in this chapter is the importance of co-ordinated design and maintenance, and of funding.

OVERCOMING FRAGMENTATION

A characteristic of modern public realm delivery is its fragmentation. There is seldom a simple single budget with a single controlling mind, or a single design team to deliver a tangible effect on the total streetscape. Public realm improvement by its nature is usually a constant act of change and improvement in which there are numerous constituent parts and activities. In the absence of a single government department being responsible for streetscape or coordinating legislation, a number of voluntary co-operation techniques help interested groups work together. Some county highway authorities have established local committees to work regularly with their district council planning authority colleagues to ensure that wider streetscape issues are addressed in road schemes. Area-wide improvements promoted by local groups or BIDs help bring together expertise and joint funding.

TECHNICAL ISSUES

In addition to the fragmentation of implementation there is a tendency for technical experts to concentrate on a specialism (Box 5.3). Most academic and professional institutions keep within strict boundaries that appear to subdivide rather than amalgamate. Often a designer will be within a contracting company or organisation, acting primarily for that company, which in turn is responsible for only part of the total streetscape. It is therefore helpful for the people making key technical decisions to understand the specialist objectives and limitations of others involved in improving the design and maintenance of a street. Many activists and campaigners champion single issues, often in isolation from others of equal importance. Sometimes an advocated course of action is over-simplified and ignores other essential priorities, leading to the activists being frustrated when their ideas are not put in place. In addition the roles of elected councillors and local Neighbourhood Plan steering groups include making important decisions on technical issues.

Box 5.3 Professions, etc. involved in street design and maintenance

Architects
Art and recreation specialists
Cleansing managers
Conservation and design specialists
Cycling experts
Disabilities advisors
Ecologists
Elected councillors
Estate surveyors
Historic building specialists
Highway engineers and managers
Transportation professionals
Landscape designers
Local amenity groups
Market managers
Members of parliament
Motoring experts
Neighbourhood plan steering groups
Police officers
Public transport operators
Recycling managers
Road safety experts
Security officers
Service company engineers
Street lighting engineers
Tree experts
Town planners
Urban designers
Walking specialists
Wildlife experts

Their professional organisations are listed on page 140

For all these reasons the programmes that disseminate specialist technical knowledge regarding the total streetscape and in a form that can be understood by non-specialists are helpful, particularly when they lead to the essential coordination of professional expertise. The problem has been recognised for many years and there are organisations that strive to liaise between professions and disciplines (Note 5.4).

Conclusion

Improvements to the public realm and streetscape are usually funded by central government from general taxation and delivered through local highway authorities as allocated budgets for highway improvements and maintenance and funds specifically targeted at objectives such as public transport, road safety, on-street parking and provision for cycling. Central funds are also directed towards improvements to listed buildings and security, and through district and town councils that have responsibilities for other matters affecting streetscape such as shop fronts, public landscape and signage.

Funds raised from the private sector that affect streetscape include payments in association with property development through the town planning process such as Section 106 planning gain and the Community Infrastructure Levy. There are also other private funds, particularly those raised through the self-imposed levies on businesses administered at Business Improvement Districts, that can make a significant impact on streetscape because of the additional leverage and co-ordinating function provided by BIDs. These voluntary payments demonstrate the economic benefits of quality streetscape, particularly in the long-term where landowners are likely to gain the most: a fact, recognised by the owners of some large freehold estates, that has resulted in the design and management of spectacular streetscapes achieved through positive estate management.

A satisfactory streetscape is often precluded by fragmentation of effort and poor coordination rather than lack of funds per se. So, in the absence of coordination enforced by legislation, voluntary arrangements are put in place led by joint council committees, BIDs, freeholders or local groups, or even a combination of all of these. To support these endeavours, the dissemination of specialist technical knowledge in a way accessible to non-specialists helps professional experts work more closely together.

Tidy integrated design includes two-stage crossings, cycle parking and landscape as well as clutter free pavements and is still well-maintained fifteen years after installation. Kensington High Street.

Part II
CASE STUDIES

The six case studies in Part II have been selected to examine how the design objectives discussed in Part I have been applied in a wide range of locations, including an international visitor destination (St Paul's Churchyard, London), a regional city centre (Gosford Street Coventry), a town centre high street (Kensington High Street, London), a suburban village centre (Poynton, Cheshire), a suburban street (Longmoor Street, Poundbury), and a rural village (Bibury, Gloucestershire). The studies therefore reflect many issues that are replicated across the country.

Some of the schemes are simple, straightforward and small interventions in the street scene. Others are more comprehensive changes brought about by the energy of locally elected representatives that required substantial funds. A common theme is that they are each the product of considerable thought and effort. As a result they offer lessons that may be useful elsewhere, especially as they have all been completed and in use for at least four years, so there has been time to assess their performance objectively.

To enable easy comparison, each case study is analysed under three headings.

Description and background: Why the location needed an overhaul, key problems that needed to be addressed and details of the designer, budget, and timescale.

What has been done? How the five key design objectives from Part I were achieved, including commentary on particular challenges.

Evaluation and discussion: The scheme's success judged against the five key objectives using available data, and how the main problems have been reduced or eradicated.

At the time of writing, all the case study sites and other places mentioned in this book could be viewed thoroughly on the Google Maps website using the Street View tool that in many locations had an archive that stretched back to approximately 2008.

The schemes at Coventry and Poynton have been studied in *Creating Better Streets: Inclusive and accessible places* published by the Chartered Institution of Highways and Transportation (CIHT), 2018 (Note Part II.1). Its conclusions are broadly similar.

"

The City of London is the foremost financial centre in the world – due in no small part, to its pre-eminent urban landscape. Our streets and public spaces contribute to the health, wellbeing and social cohesion of our society. We ensure that our built environment is of the highest quality – for commercial and community activities."

The Lord Mayor of London
Alderman the Lord Mountevans
PROJECT CLIENT

"We concentrate on delivering innovative street designs that respond to several objectives. The essential demands of the practical movement of people and goods in a heavily congested city centre is balanced with the need to create and enhance a sense of place at locations that are known throughout the world."

Iain Simmons
Assistant Director (City Transportation)
City of London Corporation
PROJECT DESIGNER

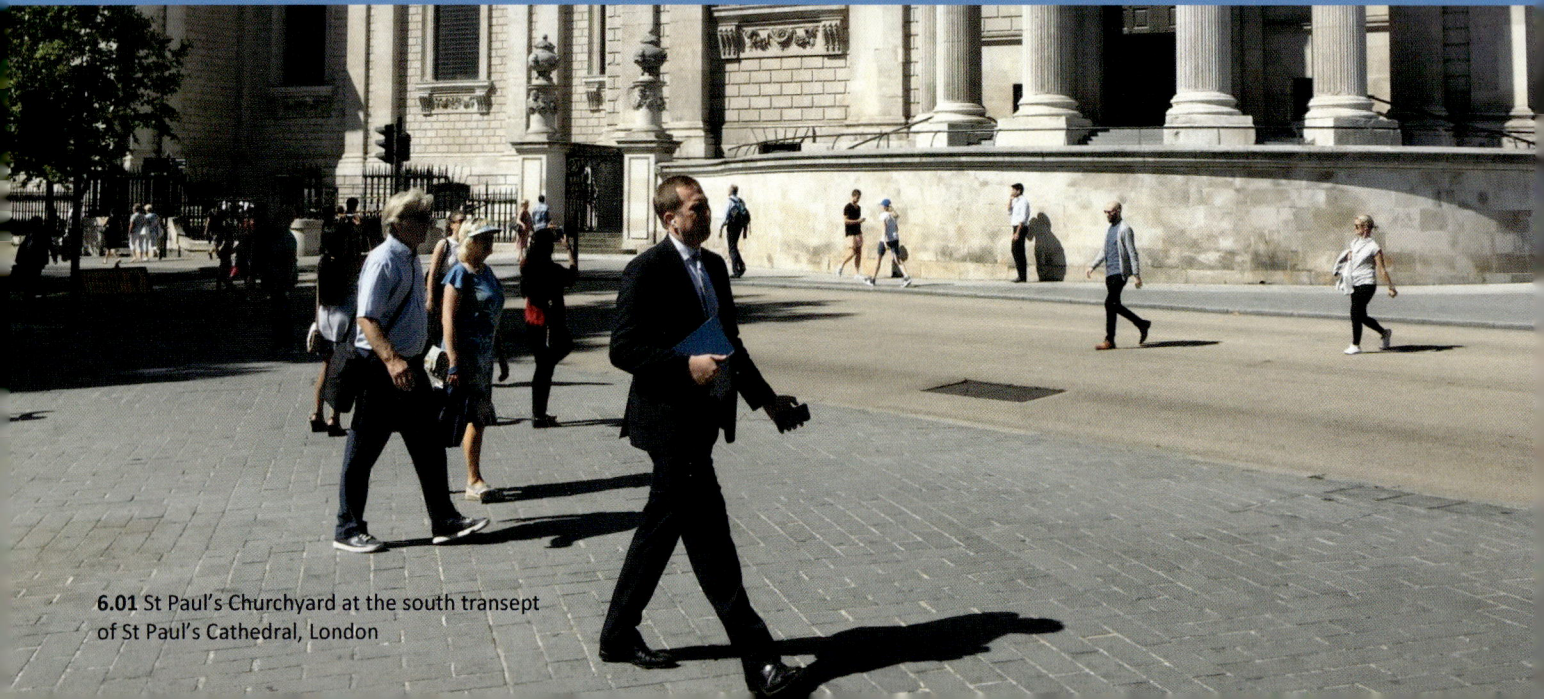

6.01 St Paul's Churchyard at the south transept of St Paul's Cathedral, London

ST PAUL'S
CHURCHYARD
CITY OF LONDON

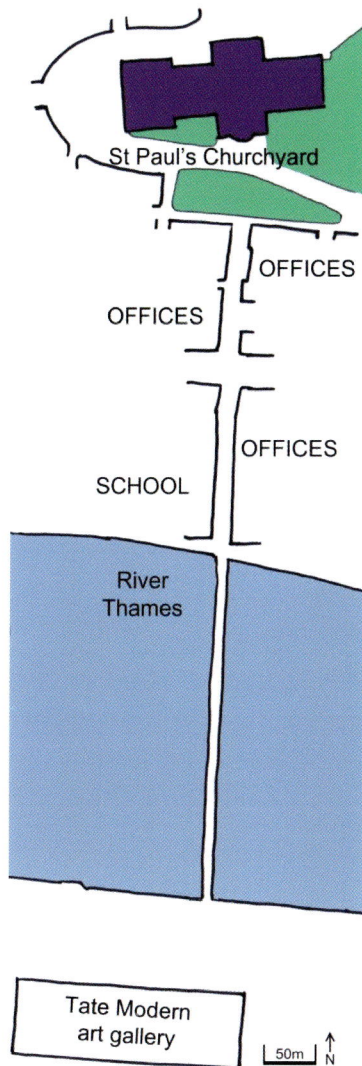

6.02 St Paul's Churchyard location plan

Description and background

NATURE AND LOCATION OF SCHEME

The scheme is a 45-metre-wide signal-controlled crossing at a site immediately to the south of St Paul's Cathedral (Fig. 6.01), built to the designs of Sir Christopher Wren in 1680 in the City of London, on the street known as St Paul's Churchyard, at the point where the new pedestrian path Peter's Hill joins it. Peter's Hill leads up from the recently constructed pedestrian Millennium Bridge that crosses the river Thames and gives visitors and tourists a pleasant walk from the attractions of the south bank of the river, including the new Tate Modern art gallery (Fig. 6.02). Even without the new bridge and art gallery, the view from the river directly to the cathedral was considered a cherished landmark purposefully provided by the construction of Peter's Hill following area-wide bomb damage in 1940.

MAIN PROBLEM AND OPPORTUNITIES

The construction of a new footbridge over the Thames for pedestrians linked two of the top twenty tourist destinations in the country by a direct, pleasant and interesting walk. The Tate Modern art gallery attracts 4.5 million visitors a year to what was before 2000 a derelict unused power station. The cathedral attracts 1.5 million visitors a year and as a result the number of people who walk between the two is increasing. This coincided with a change of planning policy by the Corporation of London, the local planning authority, that increased the priority given to the retail sector in the City and in particular at the street known as Cheapside. This catalyst for change made more urgent the need for an improved safe crossing of St Paul's Churchyard on the direct pedestrian route at the top of Peter's Hill. There was an opportunity to replace an existing signal-controlled crossing with a design that was more suitably integrated with impending improvements to the landscaped setting of the cathedral.

What has been done?

ATTRACTIVE PLACES

The crossing has been reconstructed in an adaptation of a conventional signal-controlled crossing so that although located on the pedestrian desire line at the top of Peter's Hill, none of the posts or signals of the controlled crossing can be seen to mar the distant or near view of the cathedral (Figs. 6.03 & 6.04).

The key objective of the crossing design was to ensure that people would still be able to appreciate the stunning views of the cathedral from the south side of the river and then continuously as they cross the bridge and walk in a direct line towards the famous dome of the cathedral. As they continue up Peter's Hill the dome gradually disappears from view and the colonnaded south transept of the cathedral is seen as an increasing dominant focal point of the vista all the way to St Paul's Churchyard. It was decided that although a pelican crossing was still needed to cope with both the high volumes of traffic on the road and increasing numbers of pedestrians, it should not detract from or intrude on the view.

KEY AIMS

1. Maximise the impact of the distant and near views of the cathedral
2. Integrate the design of the crossing with the new landscape at the cathedral
3. Improve visitors' experience and convenience when they use the crossing
4. Not increase traffic congestion
5. Maintain good road safety
6. Help viability at local visitor attractions

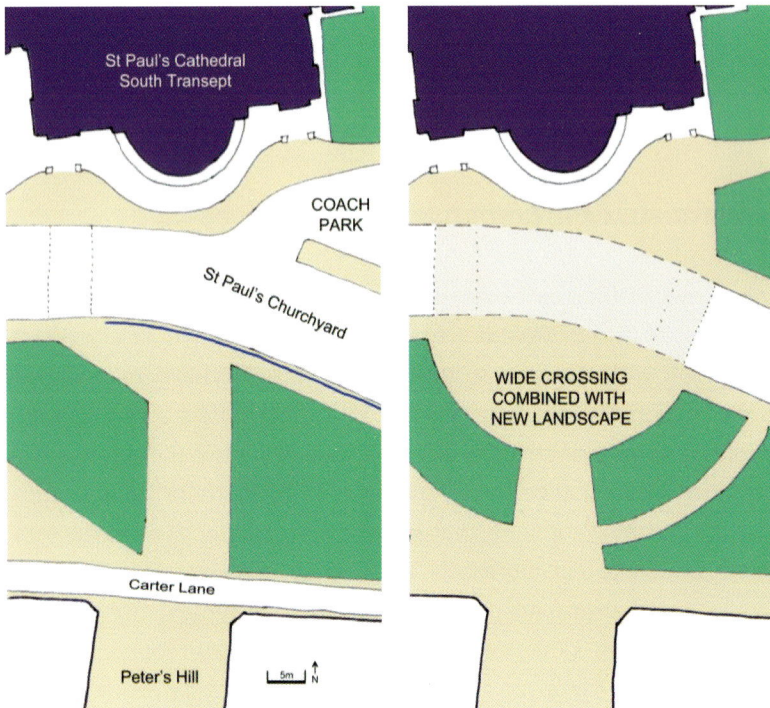

6.03 Before and after the reconstruction of landscape and pedestrian crossing

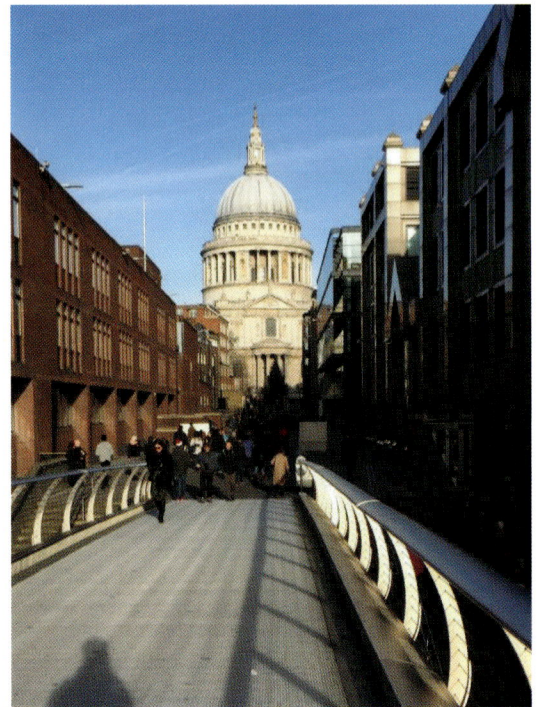

6.04 View from the Millennium Bridge

DATA	
Date completed	2012
Cost	£2m for traffic and landscape
Flows	10,000 vehicles/day (estimate)
Road safety	2005-15: 1 slight accident 2009
Designer	Highway authority
Highway authority	Corporation of London

6.05 In effect the two pelican crossings are 35m apart

The adaptation of the crossing involved two innovations:

First, the crossing is positioned precisely into a landscaped paving layout between the cathedral on the north side of the road and Peter's Hill on the south. As a result it is unusually wide. Its centre corresponds exactly with the centre point of the cathedral's colonnade and its width is determined by the width of the cathedral's south transept, a line continued at the paving edge of the Peter's Hill landscape.

Secondly the 45-metre-wide crossing is in effect two signal-controlled crossings spaced some 35 meters apart (Fig. 6.05). Each crossing is the normal 2.4 metres wide, has two sets of signals, four sets of pedestrian push buttons and double sets of studs. The signal phases of both sets of signals change at the same time in sequence, and the road surface at and between the two crossings is finished in the same unmarked plain material and colour. The appearance is intended to be that of a clear space in front of the colonnade for people to cross the road, diagonally if they wish, so that they can continue to walk to either to the right or left.

EFFICIENT MOVEMENT

The effect on traffic flow was intended to be neutral in that the crossing would not cause more congestion than before.

ROAD SAFETY

There is a speed limit on the road of 20 miles per hour. There have been no recorded accidents since the crossing was installed in 2012.

USER-FRIENDLY DESIGN

Most of the visitors who approach the cathedral from the south are on foot. Peter's Hill is quite steep and so precludes many visitors with disabilities. However, the normal tactile paving and push buttons have been installed at each of the two signalled crossing places.

FUNDING AND COMMERCIAL VIABILITY

The numbers and nature of visitors is changing. The crossing is intended to help visitors walk safely and comfortably between international visitor and retail attractions.

INTERNATIONAL VISITOR DESTINATION: ST PAUL'S CHURCHYARD, LONDON

Evaluation and discussion

ATTRACTIVE PLACES

The aims of the scheme have been met in so far as the iconic views have been enhanced. Having reached the top of the hill, people stand and gaze at the cathedral. They probably simply accept it as it is. But as international visitor attractions round the world are improved, people would notice if there was unnecessary street clutter at a cherished view.

EFFICIENT MOVEMENT

From observation, the widened crossing operates well and causes less congestion than a nearby zebra crossing on the same road where large volumes of pedestrians using the crossing seriously impede flow.

ROAD SAFETY

As expected, the road safety issues have been successfully dealt with predominantly because the layout of the crossing is straightforward. Both pedestrians and drivers can understand any dangers, and the speed limit of 20mph reduces the likelihood of severe accidents.

USER-FRIENDLY DESIGN

The concern for people with disabilities is apparent in the choice of crossing type. People with visual impairment can use the crossing unaided. If any vehicle fails to clear the road space between the two controlled crossing, it would not effect the operation of the crossing for disabled people.

FUNDING AND COMMERCIAL VIABILITY

The fact that the Corporation of London, one of the most commercially conscious local authorities, has funded and focused so much attention on the scheme is an indication that it considers the project important in underpinning the commercial viability of the area. If a similar project were undertaken elsewhere it might possibly be part funded by the local Business Improvement District (Chapter 5).

Conclusion

National crossing design regulations allow signal-controlled crossings of up to ten metres wide to be installed. The highway authority decided that a wider crossing was needed and that an adapted of two crossings was appropriate for this location. A zebra crossing might have unacceptably impeded the flow of traffic and a straight-across courtesy crossing might have unacceptably exposed pedestrians to potential danger. There may be similar locations in other towns and cities where alternatives might be considered, such as a two-stage courtesy crossing with a central refuge (Fig. 2.13). That would not impede traffic flow as much as a straight-across crossing and would reduce street clutter. However, it would not be so convenient for people with impaired vision, navigating by themselves.

As is seen at most signal-controlled crossing, the design of the conventional brackets and straps for mounting the signal heads appear to be unnecessarily visually complicated. For practical reasons the signal heads facing pedestrians across the road are usually fixed as far as possible away from the road so that a passing vehicle does not strike them. As a result, there are often long, unsightly, cantilevered brackets holding the signal heads away from the post. A practical variation reduces the brackets and another alternative reduces clutter further by mounting signals on lamp columns, as at Kensington High Street (Fig. 8.07).

The success of the scheme lies in its simplicity and the thoughtful adaptation of standard practices. The innovative use of two signal-controlled crossings placed 35 metres apart, and the careful integration of the crossing into a landscape design, enhances the enjoyment, and appreciation of a prominent building and focal point, an international visitor destination (Fig. 6.06).

6.06 The crossing enhances the enjoyment of a historic building and important focal point

INTERNATIONAL VISITOR DESTINATION: ST PAUL'S CHURCHYARD, LONDON

"

Coventry University recognises the importance of good quality public realm for its students, staff and the people of Coventry. The scheme helps open up and green our campus. It has integrated University and Coventry City Council public realm, reduced the impact of traffic and provided far more outdoor space for staff and public alike."

Professor John Latham
Vice Chancellor and CEO, Coventry University
PROJECT CLIENT

"The design was developed by our consultants Jacobs and ourselves and went through several iterations including setting out the layout with cones in a car park and driving buses round to make sure we kept the geometry as tight as possible."

Colin Knight
Assistant Director Transportation and Highways
Coventry City Council
PROJECT DESIGNER

7.01 Uncontrolled junction, Gosford Street, Coventry

GOSFORD STREET COVENTRY

CHAPTER 7

Description and background

NATURE AND LOCATION OF SCHEME

The project involved the reconstruction of a road junction and its approach roads as part of a wider improvement to the public realm at the centre of Coventry in the vicinity of the university, civic centre and the cathedral (Figs. 7.01 & 7.02). It was intended to improve the setting of the civic centre and approach to the cathedral, but particularly to complement the university's considerable development programme.

MAIN PROBLEMS AND OPPORTUNITIES

A reduction in traffic flow at the junction that had come about as a result of the completion of a nearby ring-road system encouraged drivers to take less care. This coincided with a dramatic increase in the size of the university and student numbers, and an unacceptable toll of road accidents involving student pedestrians. The university expected to continue to grow, and wished to improve the quality and extent of the public realm and improve road safety. The primary objectives therefore were a rebalancing of priorities between pedestrians and vehicles in a more attractive setting.

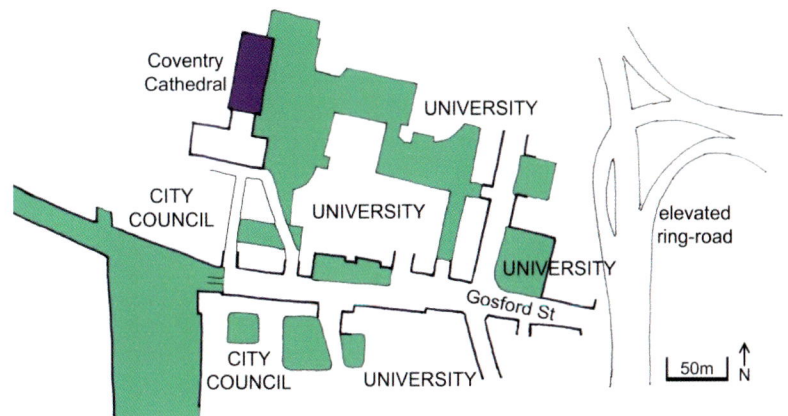

7.02 Gosford Street crossroad junction location plan

What has been done?

The two primary concerns were addressed by reconstructing the streets to give a sense of spaciousness, safety and greater priority to pedestrians. The pavements were considerably widened and the Gosford Street - Cox Street - Jordon Well crossroad, previously a conventional signalised junction, was rebuilt as an uncontrolled junction with courtesy crossings, landscape and a radical reduction of street clutter (Figs. 7.03 & 7.04). Two zebra crossings were repositioned to the east and the west of the junction. The cost of the junction and its immediate environs was some £1.1m, paid for jointly by the European Regional Development Fund and Coventry University.

ATTRACTIVE PLACES

The location is a provincial city centre, adjacent to a world-famous cathedral and within an expanding university with a growing reputation. The streetscape aim was to respect the notion of a dignified city, tempered with the feel of a welcoming and spacious centre of learning. With these notions in mind, the integrated design is intended to combine landscape and amenity features while also reducing road accidents. At its centre the crossroads looked unsightly, and was uncomfortable and dangerous for pedestrians to use; there were no dedicated pedestrian phases in the traffic signal system to help people cross any of the roads.

KEY AIMS
1. Improve the attractiveness of the city centre location
2. Create a more welcoming place for students and visitors
3. Reduce traffic speed
4. Reduce the number of road accidents
5. Reduce street clutter to a minimum

DATA

Date completed	2012
Cost	£1.1m
Flows	10,000 vehicles/day (estimate)
Road safety	2007–2011: 3 slight & 2 serious accidents. 2012–2016: 2 slight accidents (crashmap)
Designer	Highway authority and consultants Jacobs
Highway authority	Coventry City Council

7.03 Before and after reconstruction of the Gosford Steet – Cox Street crossroad junction

Each of the kerbs at the street corners had been constructed with a large radius that encouraged drivers to turn without reducing speed, adding to peoples' difficulties when attempting to cross the road.

The reconstructed crossroad and the new widened pavements were intended to give pedestrians a feeling of more space (Fig. 7.05), emphasized by each item of functional equipment in the street being visually co-ordinated and of the least intrusive design.

The rebuilt junction takes the form of a large coloured rectangle, surrounded by rows of stone setts let into the road surface and stone spheres of some 750mm diameter. Street-clutter reduction included the removal of guard railings, pavement bollards and redundant traffic signs. Mandatory traffic signs are fixed neatly to low-level stone cubes (Fig. 7.06) or existing walls. Jordan Well was reduced to a single two-way road, and its zebra crossing was repositioned in a simplified form without the two rows of guard railings. At road level the number of zigzag road markings at the zebra crossing was reduced from eight to two, and all yellow line parking signs were removed and replaced with less obtrusive Restricted Parking Zone signs fixed either to existing walls as were the Pedestrian Zone signs (Fig. 7.07) or to lamp columns. To complete the design, bus shelters, cycle racks, lamp columns and planters were replaced and incorporated into an integrated landscape scheme of new paving and street trees.

7.04 Before and after views from Gosford Street looking west across the junction towards the city centre

CASE STUDIES
REGIONAL CITY CENTRE: GOSFORD STREET, COVENTRY

EFFCIENT MOVEMENT

As the road had formerly been designed to take cross-city traffic that had been reassigned to an inner ring-road the high capacity of the road was unnecessary,so road widths were reduced from four or five to just two lanes, one in either direction. The single approach lane into the junction from each of the four directions was intended to improve the flow of traffic, as it would be easier for drivers to manoeuvre through the uncontrolled junction.

7.05 Courtesy crossing across Cox Street north of the uncontrolled junction. Pedestrians use the stone spheres for guidance

ROAD SAFETY

The slow approach speed was intended to help drivers cope with other vehicles at the junction, in the absence of signs and road markings. It was also expected to help them see and have the time and space to stop for pedestrians on the courtesy crossings. In fact pedestrians have three options for crossing the road: use one of the zebra crossings positioned some 30 metres away from the junction to the east and west, use the courtesy crossings at each of the four approach roads or cross the junction in a straight line.

USER-FRIENDLY DESIGN

The scheme is part of substantial regeneration and improvements to the quality of experience for pedestrians as they walk across the city centre. Grim pedestrian underpasses with their attendant pedestrian barrier railings have been removed and replaced with easier-to-use ground-level crossings to give pedestrians the perception that their routes across the city centre are pleasantly joined up. This extends to a new landscaped green bridge spanning an urban motorway section of the inner ring-road. Bus shelters, cycle-racks, parking places set back from the road lanes, trees and landscape have been provided within a co-ordinated design.

FUNDING AND COMMERCIAL VIABILITY

The European Regional Development Fund and Coventry University met the costs of the scheme equally. The university was keen to improve the experience and safety of students in and around the campus. The area is adjacent to the council's civic buildings and cathedral, so has an impact on civic pride and visitor experiences.

7.06 Neatly applied mandatory traffic signs

7.07 Neatly fixed zone signs

Evaluation and discussion

ATTRACTIVE PLACES

The predominance of students and the refreshingly uncluttered layout give the impression of relaxed spaciousness. When understood with the comparable upgrading of the adjacent public realm at the civic centre and cathedral, the scheme is a success because it has tackled the visual problems of a road junction that are often ignored as being impossible to improve. The complete rebuilding and the methods of clutter reduction through the adoption of a successful Restricted Parking Zone in such a large area provide useful experience applicable nationwide.

There is a point on the design of the rectangular paved area of road at the centre of the junction. Its purpose is to take away the certainty that drivers feel on a conventional road. Approaching from each direction, a driver has the sense that the road ahead is not straightforward. But the shape chosen, the rectangle, is not intended to be a new national road layout to be used as a standard alternative to a mini-roundabout. In another location the road layout might be more integrated with the layout, design and style of adjacent buildings.

EFFICIENT MOVEMENT

Because buses are no longer held up at traffic signals, their average journey times have reduced. This would also apply to all other vehicles, so the rebuilt junction is operating more efficiently than previously. The 5.5m wide roads throughout, intended to help restrain traffic speed to 20mph, appear not to impede the overall flow or efficient movement.

ROAD SAFETY

Road safety statistics show a reduction in the numbers and severity of accidents at the junction, dropping from three slight and two serious accidents in the preceding five years to two slight accidents, of which one involved a pedestrian, in the four years after completion. Removing the false certainty of traffic signals with the obvious uncertainty of an innovatively laid out uncontrolled junction and courtesy crossings would appear to be the reason for this reduction. Drivers can be seen to concentrate when they use the junction and the city council records that average speeds have been reduced from 24mph to 15mph.

Of the three optional methods to cross the road (Fig. 7.08), during our observations on-site pedestrians overwhelmingly preferred to use

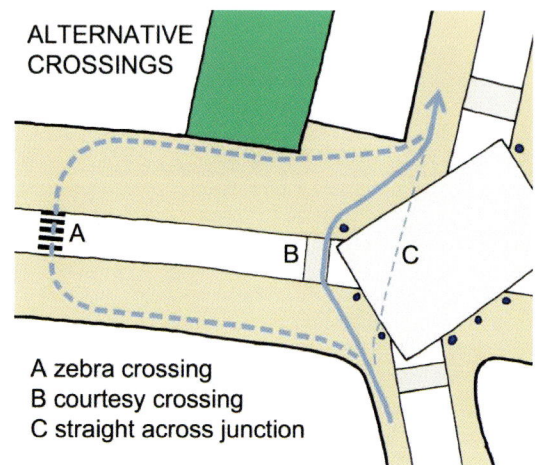

ALTERNATIVE CROSSINGS

A zebra crossing
B courtesy crossing
C straight across junction

7.08 Pedestrians have three alternative routes to take when they cross the road the courtesy crossings, and to take a straight line across an edge of the rectangular junction only when they could clearly see that there were no vehicles (Fig. 7.05). We saw no occasion when pedestrians tried to assert a right of priority to use the rectangle as they would at a Swiss 'Encounter' Zone and expect a driver to stop for them (Chapter 4). People with disabilities are helped to cross the heavier east-west traffic flows by using the two zebra crossings. But to cross the north-south flow they would need to use the courtesy crossings or seek assistance.

USER-FRIENDLY DESIGN

The provision for user-friendly design appears to be very successful, primarily because of the care taken on consultation with user groups during the design stages and the predominance of a single user type: the student. However, it is a scheme that should cater well for people visiting the city. Maintenance standards for both hard and soft landscape including street trees are reasonably high, and the obstruction-free routes along a wide pavement to friendly landmarks such as a bench or lamp-post should help visually impaired people to navigate as independently as possible.

FUNDING AND COMMERCIAL VIABILITY

The satisfaction expressed by the university is a clear confirmation that its funds have been wisely spent to provide tangible benefits for students. The city council feels the EU funding has been well spent to strengthen economic and social cohesion in Coventry.

Conclusion

The scheme takes forward at one location many themes in the practice of improving streetscape in city centres. It has overcome the practical challenges met at many urban road junctions, such as the efficient movement of traffic and road safety. This was done by using the theories of driver behaviour to create a place that gives the impression of relaxed spaciousness at an important part of a growing university campus that welcomes visitors. The scheme was funded as an essential part of the expansion of the university, important to the regeneration of the city. These techniques can be used in other urban areas. In the next chapter we examine a similar approach applied to an urban area of a very different character and with heavier traffic flows.

"We tested the legal force of some of the technical advice we had been given. Though initially advised it was legally binding we found much of it was not. Keeping fully within the law we rigorously scrutinised every concept and detail of the design. Unnecessary visual and physical clutter was removed. The viability of the high street has been uplifted and road accidents have reduced."

Councillor Daniel Moylan
Royal Borough of Kensington and Chelsea
PROJECT CLIENT

"Due to the clear-sighted, determined political leadership and progressive local authority officers we created an innovative and successful urban highway, good pedestrian circulation and road safety and a landmark high street with a distinctive sense of place."

David Moores
Technical Director, Public Realm, Project Centre Limited
PROJECT DESIGNER

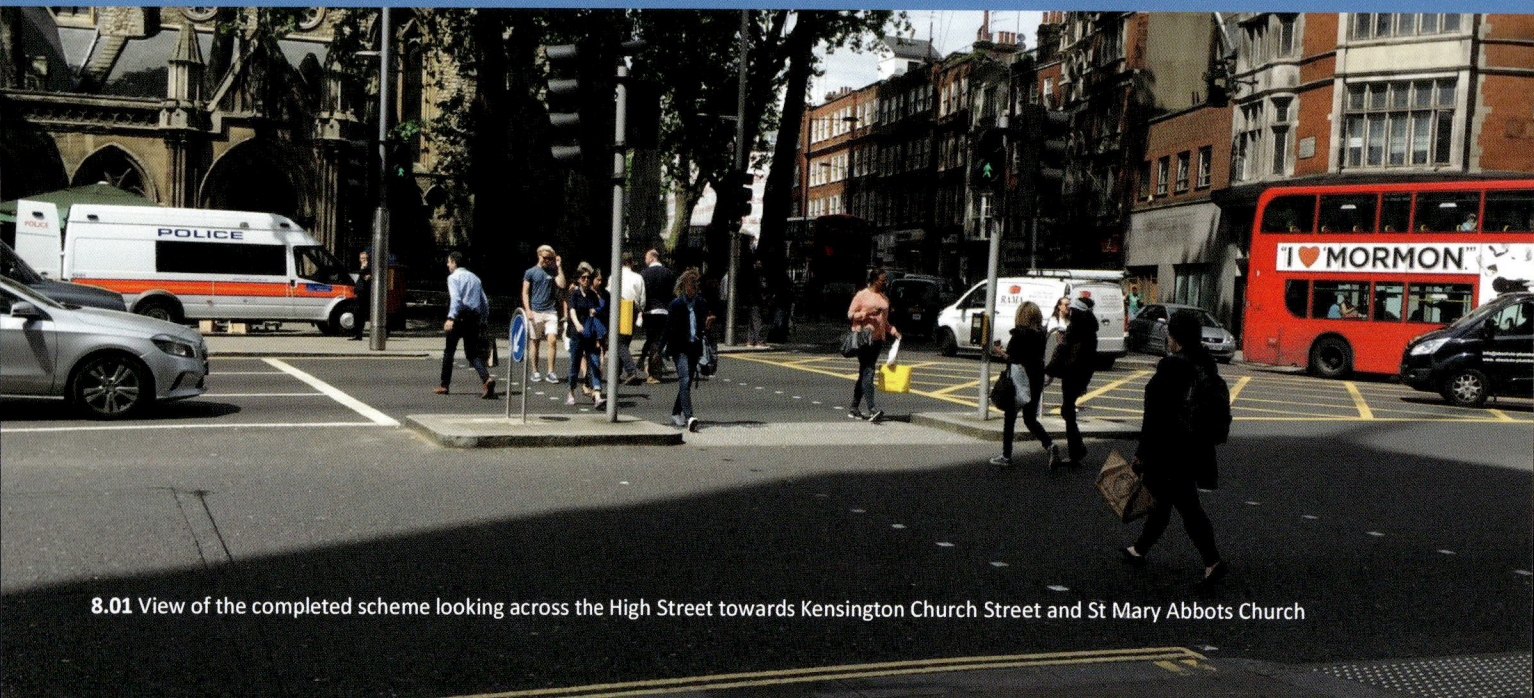

8.01 View of the completed scheme looking across the High Street towards Kensington Church Street and St Mary Abbots Church

KENSINGTON HIGH STREET LONDON

CHAPTER 8

Description and background

NATURE AND LOCATION OF SCHEME

Kensington High Street (A315) carries some 25,000 vehicles a day. The street is also functions like a typical town centre high street. It has fashion shops, national retail outlets, department stores and a busy Underground station. The unusual local characteristics are high property values and wealthy residents who expect high standards at their location of choice: an expectation fulfilled by the council's determination to retain quality in their local streetscapes.

MAIN PROBLEMS AND OPPORTUNITIES

High traffic flows have to be accepted and accommodated. The scheme centres around the junction of Kensington High Street with Kensington Church Street (A4204) (Figs. 8.01 & 8.02), and extends westwards for some 500m and eastwards for some 100m (Fig. 8.03). Although there were no undue traffic or road safety issues, the council was concerned about the depressing effect of traffic on business and the whole street.

8.02 The guard railings of the previous junction design

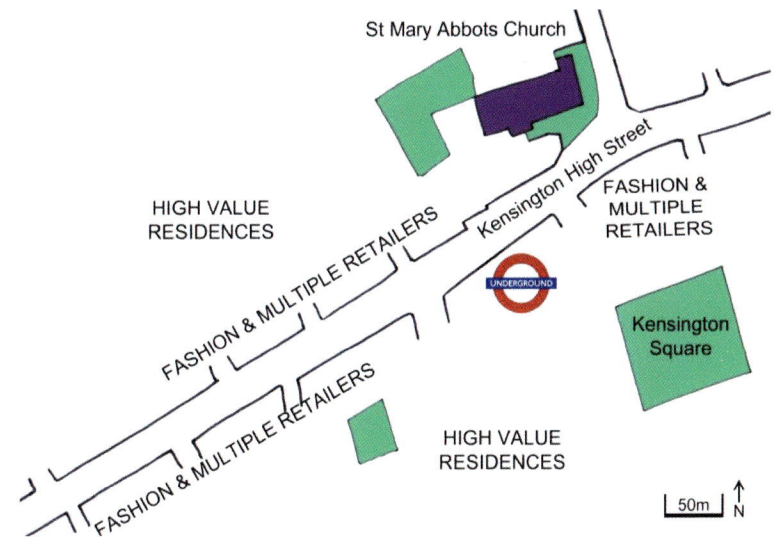

8.03 Kensington High Street location plan

What has been done?

ATTRACTIVE PLACES

The scheme came about because the borough council was convinced that the amenities of the street, especially for pedestrians and shoppers, should be radically improved. A senior councillor took the lead, and became immersed in the strategy, technical detail and implementation of the project, engaging a multi-disciplinary team consisting of private-sector consultants and in-house staff. A primary concern was to produce traffic layout designs that helped rather than hindered pedestrians as they attempted to cross the road, particularly at the major junction with Kensington Church Street. After several layouts had been examined, a simplified junction layout was agreed on that changed the signal arrangements and filter lanes, removed a triangular pedestrian refuge and, without changing the traffic flows across the junction, allowed pedestrians to cross the High Street more comfortably in a single stage instead of three stages (Fig. 8.04). At the same time some six rows of guard railings at the kerbs were removed.

KEY AIMS

1. Raise the quality of the street to a level that local residents would find acceptable.
2. Reduce the intrusive effect of traffic and improve the convenience of pedestrians particularly when crossing the road.
3. Reconstruct roads and pavements, adopting uncompromisingly clean designs, landscape and high quality materials.
4. Ensure that practical street equipment and street furniture is well designed and visually co-ordinated.
5. Eliminate unnecessary street clutter.
6. Retain existing traffic flows and not reduce road safety.
7. Retain and improve the viability and attractiveness of the street as a shopping centre.

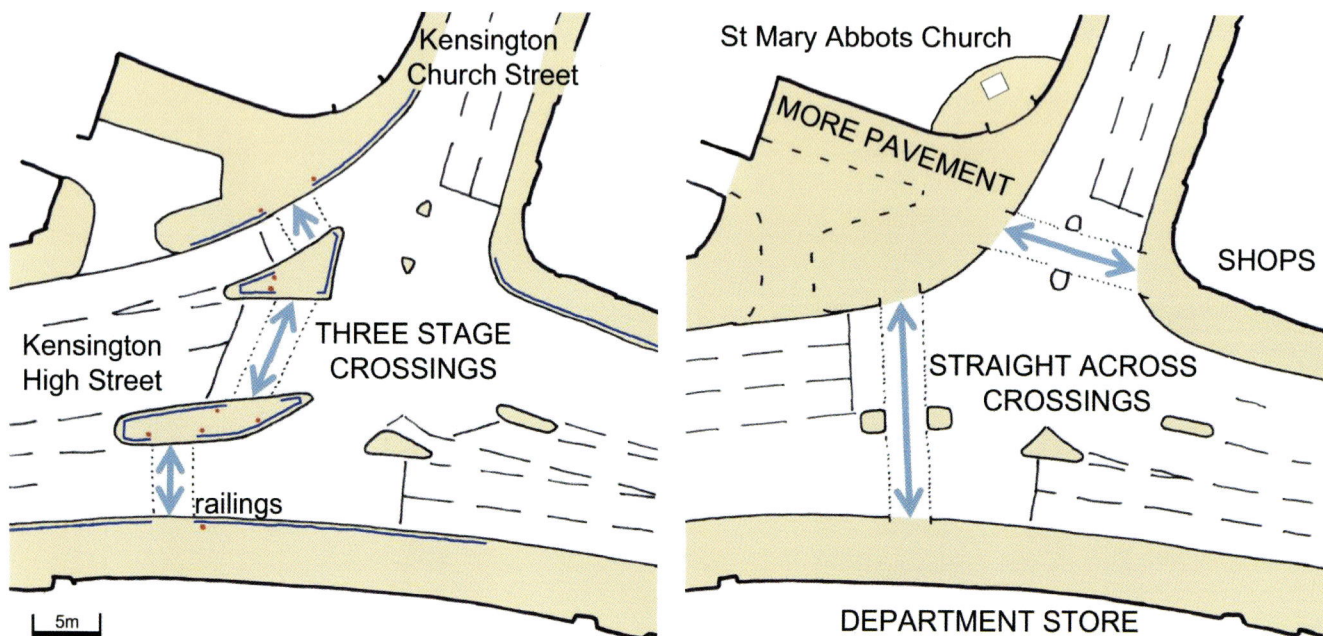

8.04 Before and after the reconstruction of the tee junction. Crossings were simplified and more pavement space was created

Away from the junction, single and two-stage signal-controlled crossings along the remaining length of the High Street were similarly redesigned and their guard railing removed (Fig. 8.05). The removal of the railing, a practice carried out previously to a lesser extent at other schemes, notably at Strand in central London, was considered dangerously controversial and attracted considerable debate. A clean uninterrupted kerb line at a road of uniform width commonly seen in central London streets was achieved by omitting raised or inset areas of road surface to provide for parking or deliveries. Along the centre of the high street there is a wide refuge, often forming part of pedestrian crossings, and including street trees and cycle parking racks (Fig. 8.06). Additional cycle parking is on the road along side streets (Fig. 4.07).

Throughout the scheme crossings, kerb lines and centre-of-road pedestrian refuges were built in unusually well co-ordinated designs using high -quality construction materials such as yorkstone paving slabs, granite setts and landscape. Though the traffic signals conform to the national design they have been assembled in a neater manner than is normal (Figs. 8.07 & 8.08). Wherever possible they are combined with street lighting columns. This involved changing the position of some lamp columns to coincide with the location of traffic signals. Special lamp columns were commissioned with twin access panels to the independent fuses, and switchgear of the street lighting and signals. The columns can also be adapted for litter bins, flower baskets and bus stop signs. The reduction of street clutter extended to limiting the number of litter bins to those on the lamp columns and removing pedestrian direction signs. This resulted in the existing pavement appearing to have been widened and be more spacious.

DATA

Date completed	2002
Cost	£4.8m (complete scheme)
Flows	25,000 vehicles/day
Road safety	2005–2015: 13 slight and 4 serious accidents at junction
Designer	Highway authority and Project Centre consultants
Highway authority	Royal Borough of Kensington & Chelsea

8.05 Guard railing was removed at the reconstructed two-stage crossings

TOWN CENTRE HIGH STREET: KENSINGTON HIGH STREET, LONDON

EFFICIENT MOVEMENT

It was intended that the level of traffic flow would be maintained; the design of the signal phasing at the junction and the two-stage pedestrian crossings in the high street assist traffic flow. Care was taken not to delay the flow of traffic and to cater for the wide range of vehicles – cars, vans, lorries, buses, taxis and cycles – that use the road. The decision to retain a traditional straight length of kerb on both sides of the road throughout the scheme in order to create a simple visual line also allowed for the flexible use of the kerb-side for deliveries, bus stops, emergencies and so on when needed.

ROAD SAFETY

The design was not specifically intended to reduce road accidents as these were recorded as being about average for the time and for the category of road junction. However, it was intended that the number of accidents would not increase.

USER-FRIENDLY DESIGN

The normal urban centre location dictated that the range of other activities along the pavements was limited. The road is, however, close to several large open spaces and it is in these that a full range of leisure activities takes place. There are no dedicated lanes for buses, taxis or cycles. Cycle parking is provided among trees planted along much of a continuous central refuge and along the carriageways of side streets.

FUNDING AND COMMERCIAL VIABILITY

A high priority was the improvement of commercial viability for the retail outlets along the street. An improvement of the shopper's experience by improving the appearance and convenience of the pavements and crossings was considered essential for the continuing commercial success of the shops.

8.06 Cycle racks are integrated with the crossings and landscape design

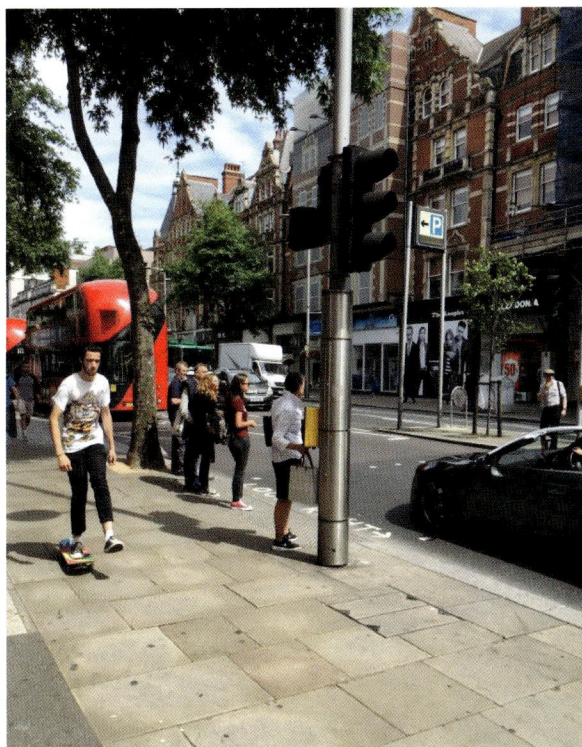

8.07 A simplified traffic light (signal) fixing is combined on a single column with street lighting

8.08 Standard fixing methods tend to look over-elaborate

Evaluation and discussion

ATTRACTIVE PLACES

When it was built at the beginning of the century, many of the design concepts in the scheme relating to public roads had not been accepted as normal practice. The removal of guard railings particularly at two-stage crossings, even though it had already been tried elsewhere, received media scrutiny and concern about road safety. However, the scheme is an undoubted success. The fact that its ideas are now becoming mainstream design practice vindicates the stance and determination of the council. It has succeeded on another level. A scheme that is some fifteen years old would normally be expected to begin to look tired, worn out or more likely reduced in quality by subsequent incongruent alterations, but this has not happened. The original concepts of the scheme are being retained, and the street furniture and paving is well maintained.

All the features of the design that are described under what has been done have proved to be suitable and successful. The reason for this probably relates to why the scheme was initially envisaged. The site is where the borough civic centre is located. The quality of the street reflects the attitudes, priorities, standards and pride of the council. Small detailed defects would be seen by council officers and councillors, and be immediately remedied; also problems would not go unnoticed by interested local residents, who are very capable of making their views known and their wishes acted upon. This demand for high standards of streetscape design quality and particularly maintenance in Kensington and Chelsea has been established for decades. It filters down and is accepted by every contractor and subcontractor. The primary lesson to gain from the scheme is that it was created and is being maintained at a very high standard because of local public demand.

EFFICIENT MOVEMENT

As was intended, the scheme has not reduced the capacity of traffic flow in the high street; it is remarkable for its ability to accommodate high flows of traffic within a civilized, thriving shopping centre. It carries similar flows to many strategic roads such, as the A20 at Dover, that cut through suburban and residential areas and require roundabouts with a large land take (Fig. 8.09).

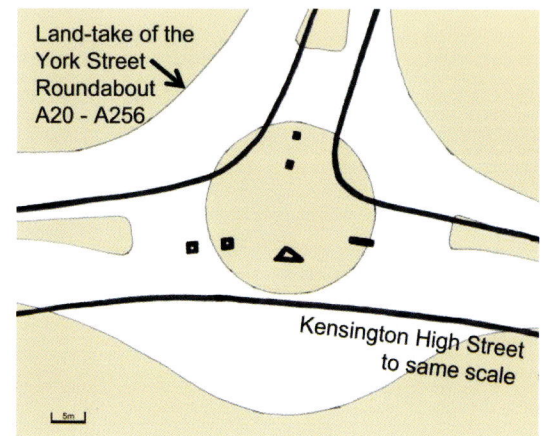

Land-take of the York Street Roundabout A20 - A256

Kensington High Street to same scale

5m

8.09 Comparison of land needed for a road and roundabout carrying similar traffic flows

ROAD SAFETY

Concern for road safety had been expressed at the design stage. It was intended that although the scheme would improve the appearance of the street, road safety would not be compromised. In fact the highway authority reports that road safety has improved. One reason is that there has been a general reduction of accidents in the UK (Fig. 3.01) since the scheme was completed. Another reason is that the crossings have a more open feel. Drivers have a clearer view of pedestrians and can anticipate their movements and because there are no guard railings drivers take more care. Since the scheme was built, designers of other projects have taken on the ideas, particularly the two-stage crossings, and have found that they can be further simplified, omitting the kerb up-stand at the central refuge without compromising safety.

USER-FRIENDLY DESIGN

The needs of cyclists, bus passengers and disabled people have all been accommodated, as previously described.

FUNDING AND COMMERCIAL VIABILITY

Kensington High Street remains an elegant, safe street where all the objectives of a complex high street are fulfilled. The higher standards of street design create safer conditions for drivers and pedestrians. The improved amenity and visual quality has helped to increase the viability of the commercial and retail businesses. This is evidenced by the continuing presence of both high-end national chain stores and small independent premium fashion shops.

Conclusion

Standards of overall design and details were painstakingly undertaken. A new suite of co-ordinated street furniture was adopted. High quality paving materials and landscape were used, and the design philosophy to eliminate street clutter on pavements such as bins, bike-racks, bollards, boxes and railing has widened the apparent pavement width and created a feel of quality and permanence. This undoubted improvement of quality has ensured the survival and viability of the street as a shopping centre. It is a credit to the local authority and its advisors and contractors, and has become a benchmark for high street design and maintenance.

"The Town Council was determined to stop the centre of Poynton becoming a ghost town: tired with empty shops and dominated by heavy traffic. We needed a radical solution. The look and the changed layout have created a thriving place. People willingly stop and chat. The traffic no longer dominates. And remarkably there are fewer and less serious road accidents."

Councillor Howard Murray
Poynton Town Council and
Cheshire East Council
PROJECT CLIENT

"Poynton has demonstrated that it is possible to create a continuous flow, low speed environment, still cope with pedestrian crossing movements and, most importantly, re-create a space, a place, outside the church that is part of the town, not an appendage to the highway."

Ben Hamilton-Baillie
Hamilton-Baillie Associates Ltd
PROJECT DESIGNER

9.01 Traffic now approaches slowly as it enters the spaciousness of Fountain Place. Vehicles arriving at St George's Church are accommodated on the highway within the scheme

FOUNTAIN PLACE
& PARK LANE
POYNTON,
CHESHIRE

9.02 Crossing the one arm of the previous junction arrangement that had signals for pedestrians

Description and background

NATURE AND LOCATION OF SCHEME

The scheme was a complete innovative redesign of the traffic arrangements, including the reconstruction of the road and pavements at a busy crossroad traffic junction located at the intersection of London Road (A523) and Park Lane/Chester Road (A5149) known as Fountain Place, in Poynton, Cheshire (Figs. 9.01 & 9.02). The scheme extends 300 metres east along the local shopping street Park Lane (Fig. 9.03) to a new supermarket that helped fund it through a planning agreement.

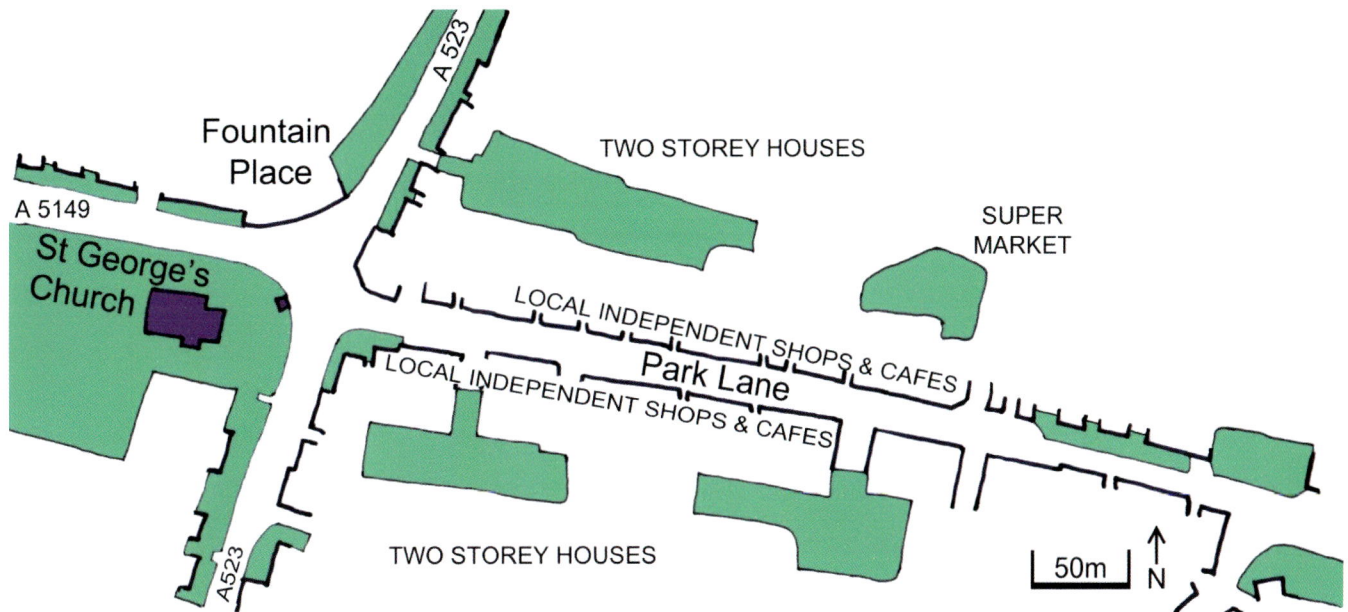

9.03 Fountain Place and Park Lane location plan

MAIN PROBLEMS AND OPPORTUNITIES

The village centre was suffering from a poor image because of heavy traffic flows and a general rundown feel, failing shops and a proposed supermarket that threatened to draw more custom away from the centre. Traffic management had been problematic for years, emphasised by heavy congestion and poor air quality, and an intersection cluttered with eleven sets of traffic signals, keep-left signs, bollards, small areas of isolated pavement and an ugly cross hatching of yellow lines. It was an intersection that was difficult for pedestrians to cross and certainly not somewhere one would choose to visit.

KEY AIMS

1. Improve the business environment and viability of the shops
2. Create a pleasant place to be recognised as the centre of the village
3. Make a 'place' with a feel of quality and well designed landscaped spacious pavements for people to visit, walk, sit and shop
4. Enhance significance of the church in the community
5. Improve the convenience for people arriving at the church by vehicle
6. Retain traffic flow levels at junction
7. Encourage drivers to interact with consideration to pedestrians
8. Allow pedestrians including those with disabilities to cross all roads safely and conveniently
9. Reduce street clutter to a minimum

9.04 The project under construction in 2010

DATA

Date completed	2011
Cost	£3m plus £1m for sewer repair
Flows	27,000 vehicles/day
Road safety	2008–2011: 7 slight & 2 serious accidents; 2012–2015: 2 slight & 1 serious accident (source Crashmap)
Designer	Highway authority and consultants Hamilton-Baillie Associates, Planit IE & Civic Engineers
Highway authority	Cheshire East Council

What has been done?

The scheme was the chance outcome of seemingly random events: the determination of the Town Council to make changes, funding through a Section 106 planning agreement associated with a new supermarket and the skills of a multi-disciplinary team experienced in less conventional traffic-related urban design projects being selected to work with the statutory highway authority. The experience of Danish, Dutch, (notably the road safety engineer Hans Monderman), German and Swiss projects is drawn on and applied to a scheme that is compliant with UK road design legislation (Fig 9.05). Though similar to the traffic scheme at Gosford Street, Coventry (Chapter 7), the roads carry heavier traffic and the local shopping centre is frequented by a wider cross-section of the population.

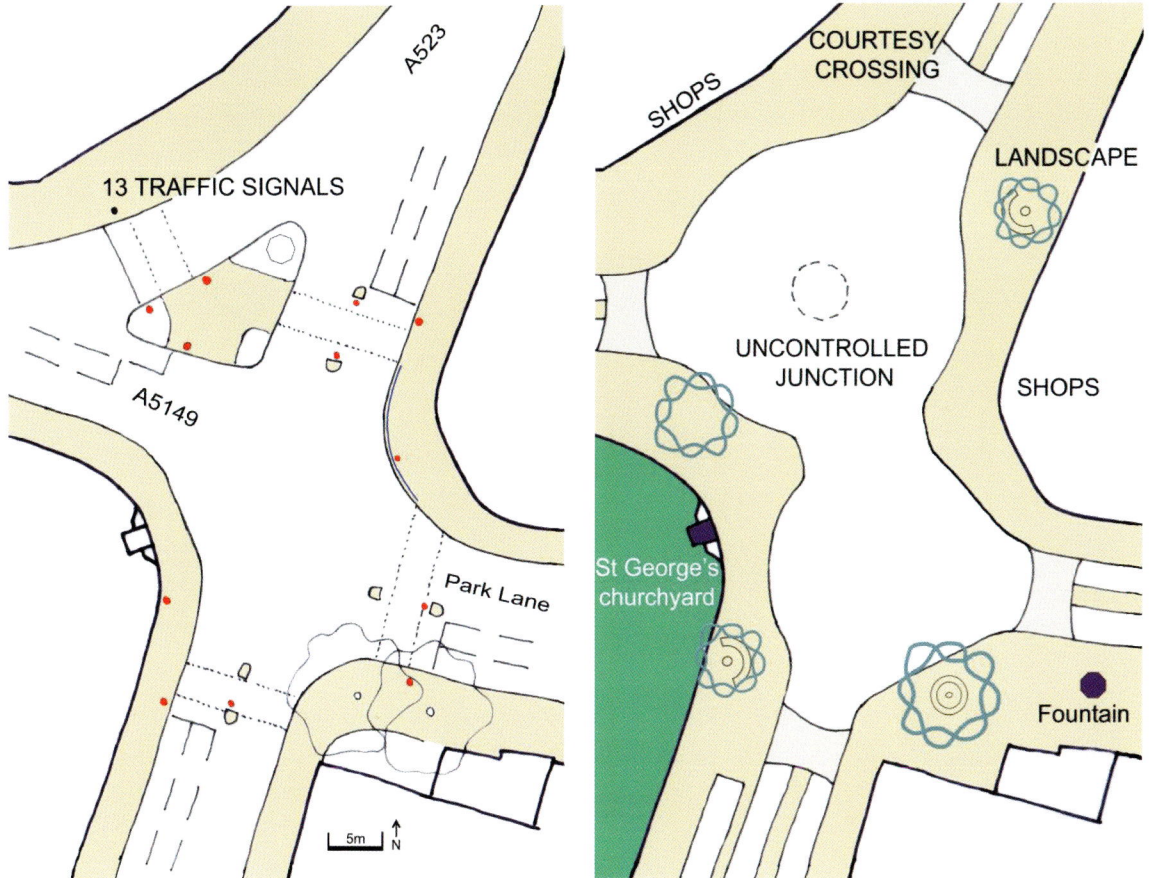

9.05 Before and after the reconstruction of Fountain Place into an uncontrolled junction

ATTRACTIVE PLACES

As discussed in Chapter 1, a starting point in improving the attractiveness of somewhere is to analyse its character as a place: *the genius of the place*, as Alexander Pope put it. Essentially the layout at Poynton is a narrow local shopping street, Park Lane, leading to a large open area that happens to be a traffic junction but is also surrounded by local shops with a historic parish church and grounds at a prominent position at one corner (Figs. 9.06 & 9.07).

9.06 Before: a large bleak expanse of tarmac

9.07 At each approach road to the uncontrolled junction there is a two-stage courtesy crossing

9.08 Before: a signal crossing, not for pedestrians

The intimate character of Park Lane is enhanced by the provision of some nine courtesy crossings, unequally spaced, on average twenty to thirty metres apart along a three hundred metre length, positioned to help shoppers cross conveniently (Figs. 9.08 & 9.09). Walking or driving from Park Lane to Fountain Place there is a sense of arrival and spaciousness, emphasised by the removal of all former street clutter, rebuilding the roads in a co-ordinated design incorporating widened pavements and crossings, lamp columns, landscape and trees.

9.09 There are nine similar courtesy crossings along the shopping street Park Lane

EFFICIENT MOVEMENT

The design objective was to retain traffic flows across a junction designed to be a shared space, defined by the scheme's designer Ben Hamilton-Baillie as *"A term, coined in 2003, to describe a wide range of principles intended to improve the relationship between people, places and traffic. Shared space builds on established social protocols and informal human negotiations, moving away from control and regulation by the state. It describes one end of a spectrum a car park, mews court or a market place, and a motorway or segregated urban highway at the other. Shared space has existed for years, and was the default arrangement prior to the introduction of formal traffic controls early in the last century. Rather than assuming priority and dominance by motor vehicles, shared space designs exploit driver behavioural psychology to extend everyday civilities to the movement and interaction of pedestrians, cyclists and drivers. Shared space is often confused with shared surfaces, where there are no kerbs or spatial divisions. Shared space is quite distinct and, as at Poynton, usually relies on clearly defined space for different users. There are few specific design elements unique to shared space. Rather the responses of drivers and reactions to other civic activity are determined by the context and circumstances."*

The single lane at each of the four approach roads allows drivers to concentrate only on what is ahead of them as they approach the junction, following a sequence of three actions as they enter and leave the junction (Fig. 9.10). They wait only a few seconds to allow pedestrians to cross the short courtesy crossing on the traffic lane on which they are driving and so they are not unduly held up.

Traffic on the junction at Fountain Place is uncontrolled and manoeuvres round two roundels bounded by raised kerbs and pavements. As at Gosford Street, Coventry (Chapter 7), no traffic management orders have been applied, so there are no traffic regulation or warning signs. Drivers treat the junction as they would a double mini-roundabout. A comparison of the conflict points at such a roundabout and at Poynton helps to explain why Fountain Place has a good safety record (Figs. 9.11 & 9.12).

In addition, drivers can cope with wedding cars or hearses that stop on the junction. They stop there until people have appropriately assembled and proceeded into the churchyard.

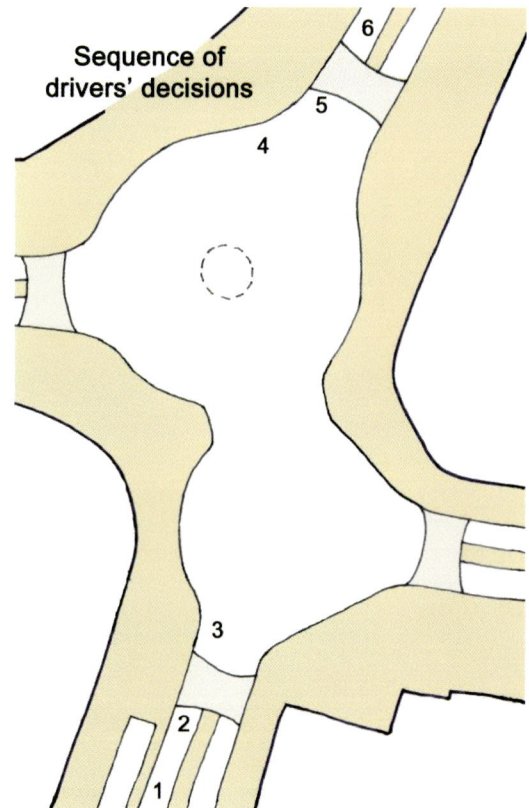

9.10 Drivers concentrate on one hazard at a time: three as they approach and three as they leave:
1. Low-speed approach and look for pedestrians
2. See and stop for pedestrians
3. Continue to junction in a single lane and concentrate on other drivers at conflict points
4. Exit junction and look for pedestrians
5. See and stop for pedestrians
6. Continue

9.11 Drivers are quite familiar with and able to cope safely at double mini-roundabouts such as at Bromham Road, A280, Bedford

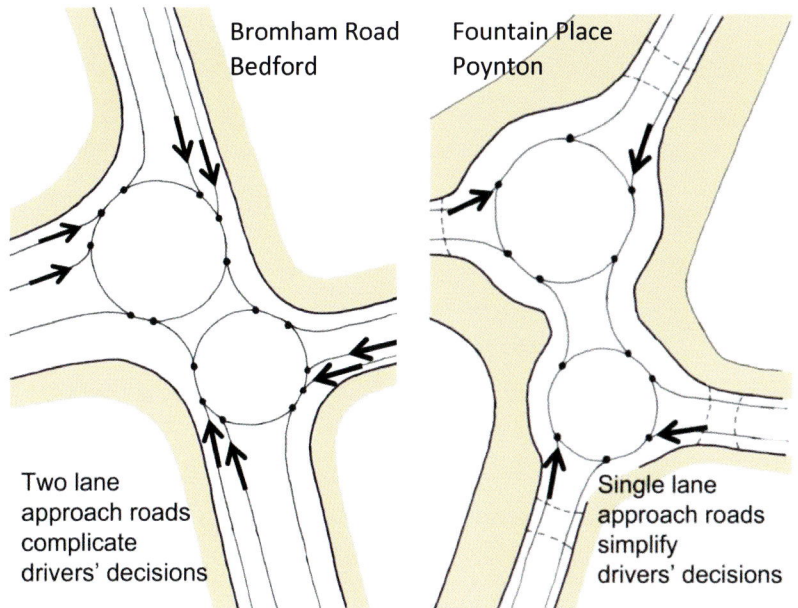

Bromham Road
Bedford

Fountain Place
Poynton

Two lane approach roads complicate drivers' decisions

Single lane approach roads simplify drivers' decisions

9.12 Comparative diagrams of the 16 conflict points at a double mini-roundabout at Bedford compared with 12 at the Poynton uncontrolled junction

ROAD SAFETY

To encourage drivers to reduce speed and be alert to the potential dangers of a road there are four groups of measures:

1. Clear transitional spaces reinforcing the boundaries between the high-speed roads outside the town and the low-speed environment of the schemes public real.

2. Edge strips in contrasting material parallel to the kerb narrow the apparent road width of roads approaching the junction (Fig. 9.13).

3. Trees, lamp columns and an emphasis on human activity provide 'edge friction' to the approaches to provide maximum interest to the drivers' peripheral vision.

4. A reduction in the linearity of the streets exploits place-making principles and emphasises spatial distinctiveness, explained further with regard to the case study at Longmoor Street (Fig.10.04).

9.13 Strips of granite setts at the edge and centre of the roads and frequent irregular courtesy crossings at varying angles to the road help reduce vehicle speed

USER-FRIENDLY DESIGN

The range of activities that take place on the pavements at Poynton is normal for a suburban village centre. The spaciousness of the design is intended to allow people to walk in comfort, and chat to passing friends wherever they happen to meet or use the benches.

The courtesy crossings have been positioned and designed to help pedestrians easily and comfortably cross each of the four approach roads at the junction. Without specific instruction, pedestrians adopt a straightforward four-stage sequence of decision-making (Fig. 9.14). They have also adopted a habit of responding to a driver that stops with a wave of the hand in acknowledgement, before and after they cross each lane of the two stage crossings:

1. Wait at the crossing until a vehicles stops.
2. Acknowledge thanks to the driver, cross to the central reservation.
3. Wait at central reservation until a vehicle in the opposite direction stops.
4. Acknowledge thanks to the driver and cross to the far side of the road.

No specific cycle lanes have been provided for cyclists as the reduced traffic speed is intended to provide a safe environment, and cycle racks have been positioned within the landscape. Public transport has not been affected and car parking and deliveries take place on the dedicated laybys within a Restricted Parking Zone.

FUNDING AND COMMERCIAL VIABILITY

The principal funding was through Highway Improvement funds, supplemented by Section 106 planning gain contributions in relation to the construction of a new supermarket in Park Lane. Additional funding came from a variety of sources, including those associated with the expansion of Manchester Airport. The prime purpose of the design was to improve the viability of businesses in the retail and hospitality sectors by building wider convenient pavements and well designed landscape to unify and enhance the shopping centre, reducing the nuisance caused by traffic and stationary vehicles at the traffic lights and providing nine new spacious crossing places between the shops in Park Lane.

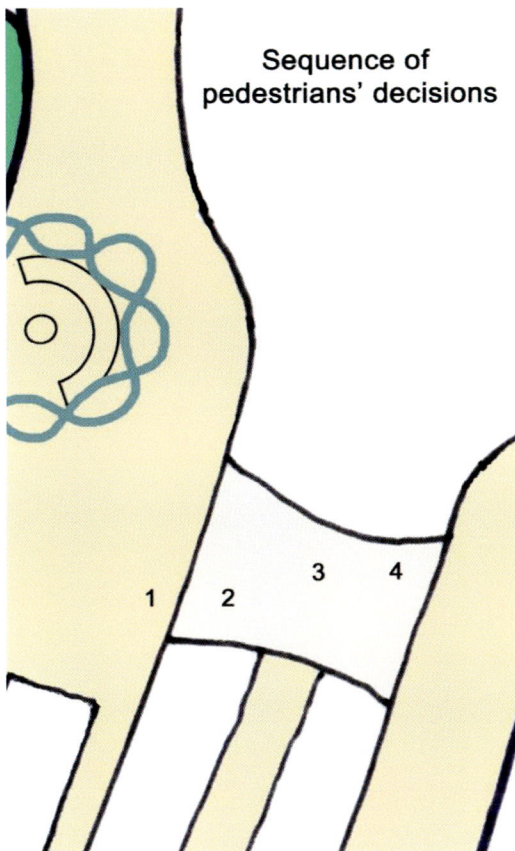

Sequence of pedestrians' decisions

9.14 Pedestrians quickly learnt to safely cope with the courtesy crossings through four simple stages:
1. Wait at the crossing until a vehicle stops
2. Acknowledge thanks to the driver and cross to the central reservation
3. Wait at central reservation until a vehicle in the opposite direction stops
4. Acknowledge thanks to driver and cross to the far side of the road

Evaluation and discussion

ATTRACTIVE PLACES

The scheme has dramatically improved the attractiveness of the area, and appears to be the reason for the change in the shopping environment and the renewed inward investment. The designers met the requirements of traffic signs regulations innovatively by adopting an uncontrolled junction. They have completely eliminated the normally imposed signs, road markings and street clutter.

EFFICIENT MOVEMENT

Total vehicular flows through the junction have not significantly changed. Drivers appear to find the even flow less frustrating than the delays at the previous signalled junction, and the shorter traffic queues in Park Lane cause fewer nuisances to shoppers.

ROAD SAFETY

The official accident data shows that the scheme has been successful. Certainly there has been no increase in the number of accidents. The data indicates a reduction in the number as well as the severity of accidents as a result of reduced vehicle speed and absence of specific warnings. Drivers understand that they need to expect the unexpected.

USER-FRIENDLY DESIGN

There is space for people to sit in comfort and it is remarkable that people meeting friends willingly stop to chat, even close to the traffic junction. The steady low speeds appear to help cyclists and the provision of tidy parking spaces adjacent to the single-lane approach roads are self-enforcing. Drivers cope with the exceptional conditions such as when vehicles attending the church stop on the junction.

The courtesy crossings operate very successfully and, as intended drivers respect and stop for pedestrians. There is a greater understanding and a more tolerant relationship between drivers and pedestrians. A comment repeated several times is that "We seem to have learnt to be nice to each other." Some dissatisfaction has been expressed by groups representing people with impaired vision, though there is evidence that some registered blind people have leant to cope safely with the courtesy crossings, possibly by using non-visual clues referred to by the chairman of the RNIB (Chapter 4), and consider the scheme an improvement.

Box 9.1

Changes to shop fronts indicating inward investment 2009-2016

(Source: Comparative visual surveys of appearance of retail shop fronts using Google Maps – Street View)

New national businesses	3
New local businesses	18
Existing businesses with refurbished shop front	15
No change	21
Vacant	2
Total shop units	59

9.15 A dreary suburban crossroad transformed

At the time of writing the issue was being addressed nationally. Possible ways forward include changes to legislation giving greater protection to disabled people as is provided by the 'Encounter Zones' of Switzerland (Chapter 2) better information and local training.

FUNDING AND COMMERCIAL VIABILITY

The scheme has proved to be an outstanding success in fulfilling its primary aim sustaining the commercial viability of local businesses. A survey of inward investment and the impact of the scheme on local property values (Box 9.1) suggest that the scheme has provided exceptionally good economic returns on the investment, and represents a satisfactory use of public and private funds, including Section 106 planning obligation contributions.

Conclusion

The scheme's primary aim of commercial viability has been fulfilled. However, it is the method by which this was carried out that is significant, and its relevance to the design of streetscapes in towns, suburbs and village centres across the country. Rather than attractiveness being seen by the designers as an objective to be balanced against an opposing objective of traffic flow, the visual attractiveness of the design has been an essential aid to traffic flow and road safety.

There have been a few similar examples elsewhere such as at Coventry, examined in Chapter 7, but the reasons for the poor take-up of the principles is that to challenge conventional design procedures and create a successful innovative scheme requires a very high technical competence, seen only in a few progressive private and public-sector organisations. In these, the multi-skilled teams are able to bring spatial design, landscape, driver behaviour, and traffic and highway engineering together to create fully integrated designs. As more schemes are produced, the advantage of a multi-disciplinary approach will be appreciated, applied not just by a team but by individual practitioners. The Poynton scheme demonstrates that the potential for improving the quality of streetscapes across the country is huge (Fig. 9.15).

"

Signs, lights and utilities. They can be easily overused. We should bury as many wires as possible and limit signage. A lesson learned from Poundbury is that it is possible to rid the street of nearly all road signs by using 'events' like a bend, square or tree every 60-80 metres, which cause drivers to slow naturally."

His Royal Highness The Prince of Wales
Point 7 of 10 Principles of Masterplanning, published by the Architectural Review, 2014
PROJECT CLIENT

"The urban layout at Poundbury deliberately reverses the dominance of vehicle geometry and instead gives a human scale and complexity to streets. By subtle alignments and gently varying street widths, speeds are calmed and the human experience of good streets emerges."

Alan Baxter
Alan Baxter Ltd
PROJECT DESIGNER

10.01 View of Longmoor Street looking west towards Burraton Square

LONGMOOR ST POUNDBURY DORSET

CHAPTER 10

Description and background

NATURE AND LOCATION OF SCHEME

Longmoor Street is in the recently built suburban village of Poundbury, at the edge of Dorchester. It is located at a point leading from what is designed as a village centre, but as Poundbury continues to expand is taking the role of a neighbourhood of Dorchester. The primary objective of the design was to create the feel of a traditional village street, with all the pleasant subtle urban design qualities that would entail. Longmoor Street was part of the first phase of construction at Poundbury and has been complete for some fifteen years, so its use has had time to settle down (Figs. 10.01, 10.02 & 10.03).

MAIN PROBLEMS AND OPPORTUNITIES

The scheme was intended from its inception to be a national design exemplar. In 1989 the Prince of Wales had published A Vision for Britain, his personal views on architecture and planning. It included ten principles that he wished to be adopted by architects and builders, and that he hoped to put into practice at Dorchester. The first three of the ten principles, The Place, Hierarchy and Scale, could be considered to be the key aims of the design of attractive places.

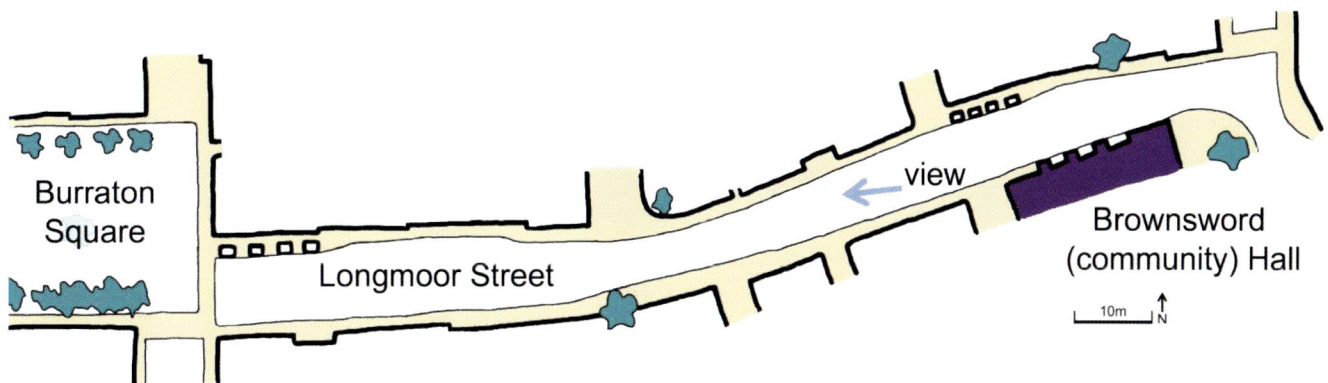

10.02 Plan showing the alignment of the buildings, pavements and road surface of Longmoor Street

What has been done?

The history of how Poundbury came to be built is well documented. It has been said to follow a familiar pattern of a joint endeavor of a royal patron working with an inspired designer. The Prince Regent and his architect John Nash created Regents Park with its iconic terraces in the early nineteenth century. Poundbury is being created on the Prince of Wales's Duchy of Cornwall land near Dorchester to the concepts of the designer Leon Krier. The original sketches show, in three dimensions, aerial views of a vision of an ideal city. Civic spaces are linked by tree-lined boulevards, houses and community buildings are arranged along a hierarchy of streets or enclosed courts, so the whole town, in this case a town extension, is composed of consciously designed interlinked spaces, each with a distinct character.

ATTRACTIVE PLACES

Possibly the most important consideration is the attractiveness of the scheme, achieved by painstaking effort and attention to the detailed design of both buildings and ground surfaces. Every view into, along and out of the street has been considered. As would be expected in a rural village, the alignment of the buildings along the street and the width of the street vary. Prominent buildings or trees close the view at each end of the street. To each side narrower streets and courts lead through enclosed squares to other parts of the village. Though totally new, the houses and community buildings are designed in classic rural Georgian and early Victorian styles. The roads and pavements are also built, as far as is practical, in a complementary style. They are constructed of simple blacktop for the main roads with bound gravel for minor roads and garage courts and pavements. There are no traffic signs or white line road markings to indicate the centre of the road, where to give way at road junctions or where or not to park.

KEY AIMS

1. Create outstanding urban design of the highest quality that would be a national exemplar
2. Use the most appropriate landscape design and materials to emphasis the character and hierarchy of spaces
3. Reduce signage to a minimum
4. Accommodate traffic acceptably and safely. Give priority to pedestrians and cyclists
5. Create a sound long-term investment for the Duchy and therefore offer practical for sale

DATA

Date completed	2000
Cost	N/A
Flows	200 vehicles/day (est.)
Road safety	No accidents
Designer	Leon Krier
	& Alan Baxter Ltd
Highway authority	Dorset County Council

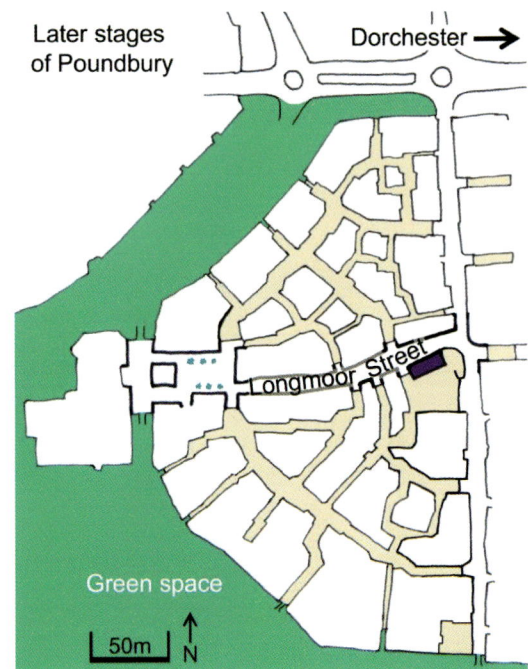

10.03 Plan showing Longmoor Street within the first stage of Poundbury

Alternative forms of streetscape

1. Buildings parallel to open ended straight road

2. Undulating building line with open ended straight road

3. As 2. with undulating road visually closed

4. As 3. with some car parking on both sides of the road

10.04 Sketch studies showing alternative forms of streetscape for Longmoor Street: (1 – 3), compared with the built version's integration of building layout with a road designed for people as well as to calm traffic (4)

EFFICIENT MOVEMENT

In a quiet residential road such as Longmoor Street, high volumes of traffic are not anticipated. Occasional large and long delivery vehicles can continue through all roads and courts without making difficult U-turns, they but are expected to move slowly and use both sides of the roads.

ROAD SAFETY

The informal layout of both the buildings and the roads together with the traditional road surfaces, and the absence of traffic signs and white lines and other road markings, are intended to encourage drivers to travel at appropriate speeds. Though within a statutory 30mph zone, there is a design speed hierarchy of 30, 20 and 10mph. Tight corners and lack of sight lines at junctions, and the lengths of a wider road for parked cars are designed to slow the traffic of speed. The sequence of sketches (Fig. 10.04) illustrate, that the layout of buildings and street can affect vehicle speed. The first shows how a straight, unending road with a white line between two parallel rows of buildings, positively encourages speed. The last of the sequence shows how the design of Longmoor Street, with its undulating road of varying widths between informal alignments of buildings, reduces traffic speed.

USER-FRIENDLY DESIGN

At the village centre there is a modest mix of local convenience shops, some specialist shops, a café, and a community hall: all the uses that would be expected at the centre of a village or local suburban neighbourhood. As there are no large traffic generators such as large car parks, non-domestic traffic includes some delivery vehicles, mixing with pedestrians. The street has been designed for the easy use of pedestrians. Crossing places have been located at the road junctions and where the short side lanes of cottages and garages meet the street. Low posts or a subtle adjustment to a kerb design indicate car parking places. Trees and benches in the small public squares at each end of the street are provided to create an informal relaxed atmosphere.

FUNDING AND COMMERCIAL VIABILITY

Poundbury was always intended to be a commercially viable long-term quality investment. It follows the examples of the private estates, mentioned in Chapter 1, stretching back to the building of Covent Garden in the 17th century.

Evaluation and discussion

ATTRACTIVE PLACES

Though the scheme has been criticised for being an architectural pastiche, many people find the architectural style attractive and agreeably consistent. The completely co-ordinated design of roads and houses that is seen at Poundbury is remarkably successful. As the buildings and ground surfaces age and mature with both the private and public landscape, the original design concept can be fully appreciated. The scheme is included as a case study because of the way in which buildings and roads have been created as a single integrated design. Though there have been many attempted copies, few have so skilfully combined the concept of a road and landscape in relation to the buildings to create a suburb that is so pleasant, agreeable and salubrious, as John Burns would have put it (page 13).

EFFICIENT MOVEMENT

A small residential street such as this would rarely pose concerns about the efficient movement for traffic unless it happened to be needed for larger flows of passing traffic. This does not occur and the arrangements for occasional larger vehicles do not appear to have caused concern. The absence of culls-de-sac helps drivers of commercial delivery vehicles to navigate into and out of the street.

ROAD SAFETY

The main lesson to gain from the design of Longmoor Street is the success of the traffic calming effect of the road's layout in relation to the adjacent buildings, landscape and public spaces. Its significance is that although it could be adopted in new developments it has many applications as a source of ideas for the innovative design of traffic calming schemes. We are accustomed to accept that traffic calming can be only achieved by legislation through 20 miles per hour zones or by incredibly ugly retrofitted road humps, concrete chicanes, rows of brash bollards or a plethora of signs and posts, or an assembly of all. Longmore Street demonstrates that if drivers are uncertain because of the lack of traffic signs of any description, coupled with a restriction of forward vision in a street and also at street corners, they reduce speed. There are more examples in Chapter 2 (Figs. 2.21, 2.22 & 2.23) and the road safety principles have been explained in Chapter 3.

USER-FRIENDLY DESIGN

Concern for the wellbeing and comfort of pedestrians has been successfully put into practice. The street has clearly not been designed to allow cars to dominate, yet car parking places have been included and integrated with landscape and street trees.

FUNDING AND COMMERCIAL VIABILITY

Commercial success is evident by the increased extent of the expansion of Poundbury. Work started on Phase 2 in 2003, and by 2013 there were approximately 2,250 people living in Poundbury and 1,660 employed in 40 businesses. The whole project is expected to be completed by 2025.

Conclusion

The completely co-ordinated design of roads as well as houses that is seen at Poundbury is remarkably successful. Walking along Longmoor Street one has the feel of a relaxed traditional village street. In these terms the scheme has achieved exactly what was intended: a national exemplar. A small residential street such as this would rarely pose concerns about the efficient movement for traffic. The main lesson to gain from the design of the street is the success of the traffic calming effect of the road's layout in relation to the adjacent buildings, landscape and public spaces. Crossing places for pedestrians have been located at the road junctions, and where the short side lanes of cottages and garages meet the street. Details such as the traditional sharp corners at road junctions give pedestrians greater sense that they have priority and the reduced sight lines help reduce vehicle speed. Though some ideas from the scheme have been copied many times in recent housing development, few if any have replicated the total integration of building style and layout with the subtle details of the design of the roads and landscape. It is significant in that although it could be more fully adopted in new developments, it has many applications as a source of ideas for the innovative design of traffic calming schemes, possibly as part of the new neighbourhood plans, in villages, small towns and suburbs across the country.

"Bibury is world renowned for its beauty and its association with William Morris. It is one of the many attractive historic towns and villages we have throughout the county. Our challenge is to improve road safety, whilst maintaining the scenic and historic character of these places for the long term benefit of our residents and visitors."

Councillor Mark Hawthorne
Leader, Gloucester County Council
PROJECT CLIENT

"This would seem to be just another road safety scheme. But instead of increasing the number of conventional road safety features, such as additional warning to drivers, we took away some fifty traffic related signs. By encouraging drivers to take more care the historic rural village is safer and certainly less cluttered."

Scott Tompkins
Lead Commissioner for Transport, Gloucestershire County Council
PROJECT DESIGNER

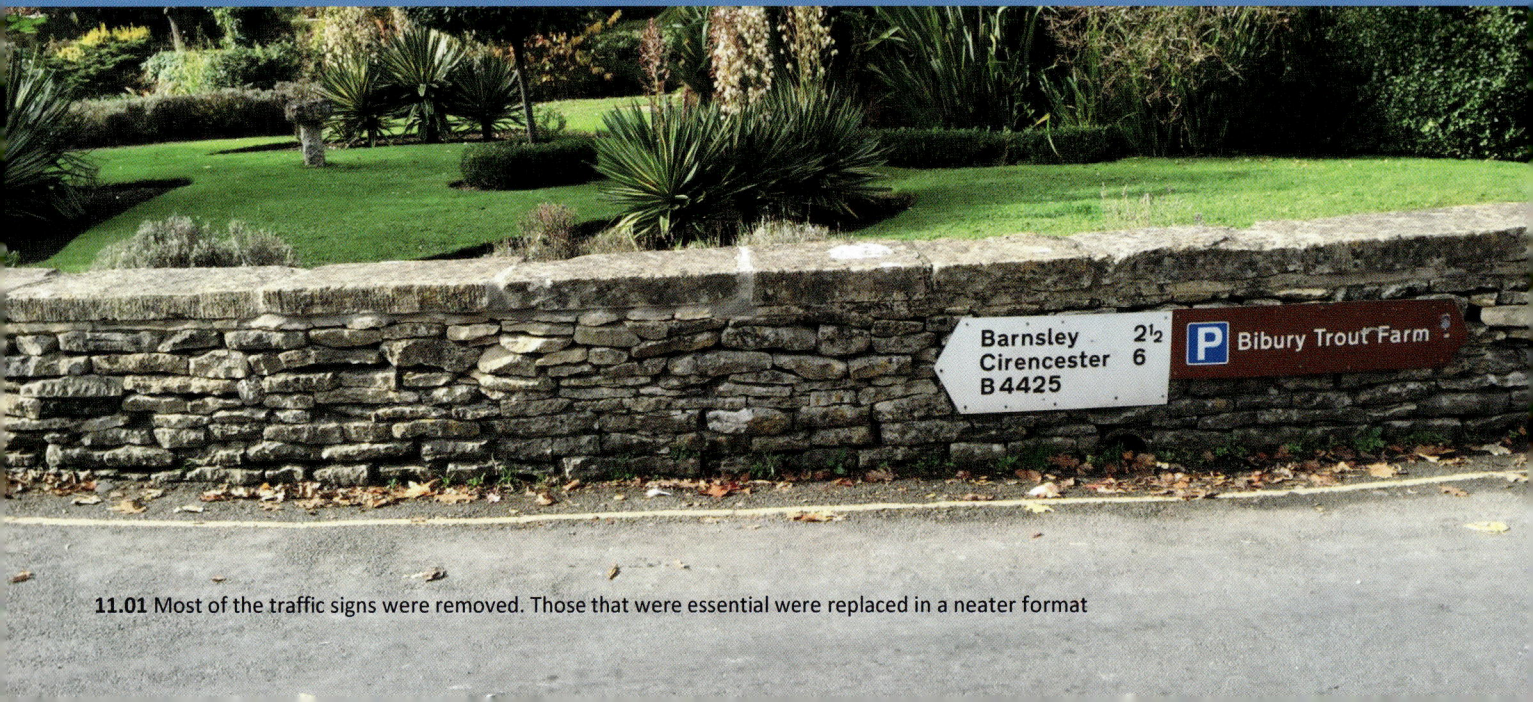

11.01 Most of the traffic signs were removed. Those that were essential were replaced in a neater format

BIBURY
GLOUCESTERSHIRE

CHAPTER 11

Description and background

NATURE AND LOCATION OF SCHEME

"The most beautiful village in England" was the how the Arts and Crafts designer William Morris described Bibury. Situated to the south of the Cotswold Hills in Gloucestershire on the River Coln, the village attracts coachloads of international visitors who come to see its 14th century stone cottages, part of a spacious historic village clustered around the Saxon parish church of St Mary and set among water meadows flanked by wooded hills. This picture perfect corner of the Cotswolds is designated a conservation area, and welcomes visitors to its former corn mill that now houses a folk and agricultural museum, herb garden and gift shop (Figs. 11.01 & 11.02).

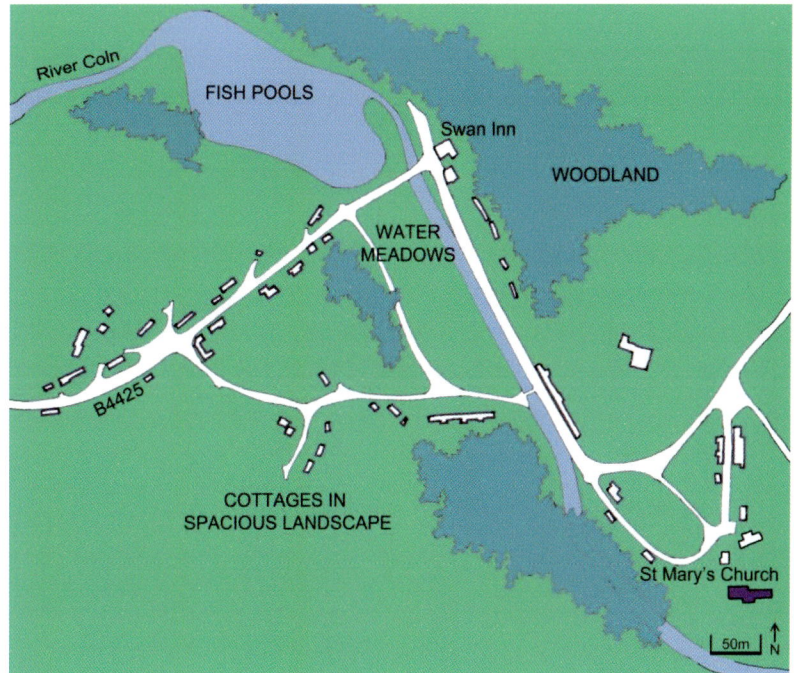

11.02 Location plan of Bibury village with its spaciousness and water meadows

MAIN PROBLEMS AND OPPORTUNITIES

Though high on the list of places to visit for international tourists, the village suffered from an unexpected number of road accidents and a glut of traffic signs that marred its beauty. Having been designated a conservation area in the 1980s, great care had been taken to ensure that the buildings and spaces of the village were protected and correctly cared for through the strict control of new development using the town and country planning acts. However, many traffic signs had been erected, at different times, in order to manage the high number of visitors and vehicles. This accumulation of signs was the subject of a conservation appraisal report by the Cotswold District Council planning authority, recommending that Gloucestershire County Council highway authority should consider reassessing the need for so many signs (Fig. 11.03). This reassessment was carried out and as a result in order to address the two problems of safety and aesthetics the scheme described here was put in place in 2005. Subsequently there have been some minor alterations that continue the same theme.

KEY AIMS
1. Restore the village to its former quality by removing unnecessary traffic signs
2. Reduce the number of road accidents

DATA

Date completed	2005
Cost	Not known
Flows	7,000 vehicles per day (estimate)
Road safety	2001–2004: 2 slight & 1 serious accidents; 2005–2008 1 slight accident (Crashmap)
Designer	Highway authority
Highway authority	Gloucestershire C. C.

What has been done?

A scheme to reduce traffic accidents was put in place in 2005. It included removing some fifty traffic signs and the subtle reduction of road widths at key locations. This reduced the speed of traffic and the number of road accidents, and at the same time improved the appearance of the village.

ATTRACTIVE PLACES

The character of the place is easy to categorise using the definitions discussed in Chapter 1. It is undoubtedly recognised by very many people as a rural village in inspiring countryside. The streetscape objective therefore is straightforward: to achieve countryside that is truly beautiful, with views both distant and immediate unmarred by clutter or obstructions. The challenge was to improve the visual qualities of the village while at the same time achieving the road safety objectives. In addition to removing unnecessary signs, the road was subtly narrowed by extending grass verges in a way that appeared to be part of the historic undulating nature of the road.

Some direction signs were retained, reduced in size and fixed neatly to boundary walls at the edge of the road, and an ugly chevron at a potentially dangerous bend was replaced with robust timber posts, each with a red reflector. Centre-of-road white lines have been removed and the effective width of the road has been reduced by parked cars (Figs. 11.04, 11.05, 11.06 & 11.07).

EFFICIENT MOVEMENT

The aim was to maintain efficient movement of traffic by reducing speed and encouraging more calmed steady flows so there are no extended lengths of road that would prevent two-way traffic.

ROAD SAFETY

Most of the road accidents had occurred because drivers did not expect people to walk in the road and most of those who walked in the road were visitors who did not expect fast moving vehicles to be travelling through the village. Lowering the traffic speed was therefore essential.

USER-FRIENDLY DESIGN

Unfortunately in most rural villages it is not possible to provide fully for all user groups and maintain a rural character. Wide, even pavements on both sides of a rural road change the character of the road to the extent that it would no longer have a rural character. As a compromise there are sufficient clearly defined pavements to cater safely for the majority of visitors. Apart from that, the slower traffic speed was intended to help cyclists and disabled people. There are no designed shared surfaces but in the centre of the village where people congregate, pedestrians do walk in the road. The village is not a densely used commercial or retail area.

FUNDING AND COMMERCIAL VIABILITY

Funding was predominantly from the road safety budgets of the county highway authority but the commercial advantages are tangible. Tourism accounts for 17% of all local employment and annual direct spend by visitors was £328m in 2014 (Note 11.1). International tourism is very significant to the economy of the Cotswold communities and to the country, and the visual quality of the countryside and its rural villages is an essential part of that commercial offer. Growing visitor numbers and improved road safety are clear indications of success.

11.03 The clutter of traffic signs was considered to reduce the attractiveness of the village

11.04 Some 50 signs were removed as being unnecessary or simply redundant

RURAL VILLAGE: BIBURY, GLOUCESTERSHIRE

Evaluation and discussion

This traffic calming scheme was one of the first to use the knowledge of driver behaviour outlined in Chapter 3. It does not rely on drivers being instructed by traffic signs, rather by changing the nature and feel of the village drivers understand that they should reduce speed, and they do.

ATTRACTIVE PLACES

The success of a traffic calming scheme is that it should be virtually invisible in the sense that the works that have been carried out should not be obvious. Certainly few people would notice that a sign was missing or that a grass verge has been extended in order to narrow a road and reduce traffic speed. These techniques could be applied to many other retrofitted traffic calming projects. New procedures in Norfolk by the county highway authority (Chapter 3) remove redundant traffic signs as road maintenance is carried out. The long-term cost reductions are obvious as in the first two years some six hundred redundant signs were removed. In Surrey de-cluttering programmes are led, in association with the county highway authority, by local parish and town councils.

EFFICIENT MOVEMENT

No concerns have been expressed about any additional impediment to traffic movement since the scheme has been in place. Rural lanes and villages by their nature restrict movement to an extent.

ROAD SAFETY

The improvement to road safety is a result of drivers being more aware that the road ahead may not be clear at all times and that they should expect the possibility that people may be walking in the road. This in addition to slower traffic speed, helped drivers to cope safely when they did encounter a pedestrian walking in the road.

USER FRIENDLY DESIGN

There have been no major changes to the provision for the wider user groups. As experienced in many rural villages, there need to be some compromises, and at Bibury there are sufficient clearly defined pavements to cater safely for the majority of visitors. Slower traffic speed has helped cyclists and disabled people. There are no designed shared surfaces, but in the centre of the village where people congregate, there are places where visitors need to walk in the road to get on to a coach.

11.05 Ugly chevron warning signs were replaced with rural-style wooden posts with reflective discs

11.06 New subtle grass verges narrowed a section of the road, fitted seamlessly into the appearance of the village and helped reduce traffic speed

FUNDING AND COMMERCIAL VIABILITY

Bibury is on the south-east edge of Cotswold Area of Outstanding Natural Beauty that attracts 38 million visitors a year and has a population of 139,000. The importance of the tourism industry is fully understood by the county and district councils, with sustainability high on the agenda. A Cotswold Sustainable Tourism Partnership was set up in 2010 to co-ordinate development, management and promotion.

Conclusions

Two lessons for general application are that subtle traffic calming using landscape and grass verges to narrow roads are effective in reducing traffic speed, and that the opportunity to reduce street clutter specifically by removing traffic signs is very considerable. Since the scheme was introduced, other traffic calming projects have been put in place in other places using similar techniques. However, by removing so many traffic signs in one short length of street, the scheme clearly demonstrates that there are probably thousands of redundant traffic signs across the country. Though the frequent response to an accident is to demand more traffic signs, crash data often show that accidents continue to occur, despite additional signs. This message is gaining strength and programmes to de-clutter are increasing largely due to the experience at a few towns and villages such as Bibury.

Large numbers of visitors continue to gain a greater understanding and appreciation of the qualities in Bibury to which William Morris drew attention. Though the roads now have to cope with modern traffic, the appearance of the village that Morris knew has to a large extent been restored. The scheme has certainly secured its future as an international a visitor destination.

11.07 Parked cars that effectively narrow the road and the removal of road centre white lines reduced traffic speed

CONCLUSION

The design and maintenance of our streets affects us all. It affects the places where we live and where we visit every day of our lives. It is an important element in the enjoyment of life. The quality of a street is also a reflection of a civilised society and, despite the frustrations caused by the fragmentation of funding, decision-making and timing of most public realm projects, is worthy of more attention.

In this book we have tried to demonstrate that streets can be attractive places, able to provide for efficient and safe movement of traffic as well as all the other uses that are demanded today. With careful design and maintenance expertise streets can fulfil all these design objectives at the same time, and still be commercially viable and attract private-sector funding.

We have analysed what makes a street attractive, and how spaces can be created with roads and buildings. We have also suggested that every street is a place that should be respected and enhanced. The technology and principles behind traffic movement have been explained, along with the growing interest in less-regulated traffic with its opportunities for street designs that relate more precisely to their context. This has been taken forward to the chapter on road safety, where an understanding of how drivers safely respond to their immediate environment can help design streets that are attractive, efficient and safe.

At the same time modern streets are required more than ever before to provide for a far wider range of business and leisure uses and activities – and for a range of users, including those with disabilities. Yet the examples in the case studies show this can be achieved. All of this also makes economic sense to the extent that private businesses are willing to contribute additional funds towards it.

THE CASE STUDIES

The case studies were selected because they include ideas and practical innovations that have stood the test of time, and so can be adapted and applied more widely in a variety of locations across the country.

St Paul's Churchyard is the lawful adaptation of the national traffic signs regulations that turned a humble signal-controlled crossing into a dramatic spatial climax to long cherished vista of a national landmark. Landmarks in other cities and towns could benefit similarly.

Gosford Street, Coventry, an innovatively designed uncontrolled junction, demonstrated that road space designed to act in a similar way to a mini-roundabout need not be round and can be integrated into the wider landscape of the street. Its success as a public street within a university district suggests that the principles could be adapted for other locations.

Kensington High Street has a high traffic flow that severs a busy shopping destination: conditions that are similar to many town and village high streets. The lessons here are the importance not only of the initial co-ordinated high-quality street design but also its subsequent continual maintenance and management, so that at the time of writing some fifteen years later its quality had not been eroded. This required constant vigilance by the local community and its local authority – a useful example for others.

Poynton is well known to professional street designers as an unconventional design that handles high traffic flows. Yet it has had few followers, partly because its design is the result of what is still in the UK an unconventional approach. Our case study assessment, coupled with the explanations of the underlying design objectives in the Chapters of Part 1, is intended to help demystify its rationale for an interdisciplinary professional audience. If applied more widely, the impact across the country could be very significant because it demonstrates that every ordinary street and traffic junction truly has the potential to be welcoming and a place one would really want to visit.

Poundbury, like all the other case studies, is the result of a strong, determined intention to create an exemplar. Our analytical diagrams compare the effect of the alignment of road and buildings on traffic speed, and how at Poundbury traffic calming devices have been incorporated seamlessly into the layout, geometry and detailed design of the street, principles that could applied more widely.

Bibury, though a modest scheme, was one of the first examples of concentrated de-cluttering as part of a road safety project. The approach is being adopted elsewhere and there is a huge potential to make a real difference quite quickly at places where de-cluttering ideas might be initiated by local residents groups.

COMBINED EFFORT AND COORDINATION

Most of the case studies were financed through joint funding from more that one source: private as well as public sector. Although work on the highway will be carried out through the local highway authority, it can be initiated by people interested in improving the places where they live, possibly by working in local groups, by local companies through their Business Improvement Districts and by large freeholders as we examined in Chapters 1 & 5. Good streetscapes make commercial sense.

We have sought to explain how the complexities can be unravelled so that people who may at present be involved only on the periphery might be more engaged and take part in the knowledge that they can make a real difference. And for people who have a professional interest in design, we hope that the book might help fill any gaps in their knowledge of some technical aspects, perhaps on road safety or funding systems, and so make more effective their endeavours to design and deliver great streets.

NOTES

Chapter 2

2.1. Page 27
RAC statistics on journey to work
The car and commute. RAC Foundation Fig 1 Table 2
http://www.racfoundation.org/assets/rac_foundation/content/downloadables/car-and-the-commute-web-version.pdf
Journeys to central London by car. Table 2.5
http://content.tfl.gov.uk/travel-in-london-report-8.pdf

2.2 Page 31
Transport for London study on safety purpose of guard railing
http://content.tfl.gov.uk/guidance-on-assessment-of-pedestrian-guardrail.pdf

2.3 Page 32
Design and layout of signs and signals at junctions and crossings
The Traffic Signs Regulations and General Directions 2016
http://www.legislation.gov.uk/uksi/2016/362/schedule/14/made

2.4 Page 36
Department for Transport. Local Transport Note 1/07 & Highways (Traffic Calming) Regulations 1999
https://www.gov.uk/government/uploads/system/uploads/attachment_data/file/329454/ltn-1-07_Traffic-calming.pdf.
http://www.joincrash.com/files/Traffic%20Calming%20Regulations%201999-1026.pdf

2.5 Page 38
Quiet Lanes and Home Zones (England) Regulations 2006
http://www.legislation.gov.uk/uksi/2006/2082/pdfs/uksiem_20062082_en.pdf

Chapter 3

3.1. Page 43
Overview of road accident data
http://www.gov.uk/government/statistics/reported-road-casualties-in-great-britain-main-results-2015

3.2. Page 43
Plane crash data
www.planecrashinfo.com/2015/2015.htm

3.3. Page 43
Road accident data, some details
https://www.gov.uk/government/uploads/system/uploads/attachment_data/file/648081/rrcgb2016-01.pdf
See Tables RAS 30005. Reported killed or seriously injured casualties, by road user type, Great Britain, 2005 – 2015 & RAS 30022. Reported killed or seriously injured by day, road user and hour of day Great Britain 2015

3.4 Page 47
Studies into driver inattention and distraction in relation to crashes.
100-car Naturalistic Driving Study – USA Naturalistic Driving Study. The Impact of driver inattention on near-crash/crash risk 2006 & Exploring inattention and distraction in the SafetyNet Accident Causation Database – Talbot, Fagerlind & Morris UK 2013

3.5 Page 47
Typical car stopping distances
www.gov.uk/browse/driving/highway-code-road-safety
Rule 126.
https://assets.publishing.service.gov.uk/media/559afb11ed915d1595000017/the-highway-code-typical-stopping-distances.pdf

3.6 Page 52
Study of less regulated traffic scheme at Bexleyheath
Designers: London Borough of Bexley and Phil Jones Associates
http://discovery.ucl.ac.uk/1503196/3/Jones_TPM%202015%20Horrell%20Final.pdf

3.7 Page 52
Need for traffic warning signs
Gorringe v Calderdale MBC. House of Lords 2004
http://www.publications.parliament.uk/pa/ld200304/ldjudgmt/jd040401/gorr-1.htm

3.8 Page 53
Quality audits
Traffic Advisory Note 5/11 Quality Audit
https://www.gov.uk/government/uploads/system/uploads/attachment_data/file/4394/5-11.pdf

3.9 Page 54
Accident data by location
Crash map website
http://www.crashmap.co.uk/search

3.10 Page 54
Main road traffic count data
https://www.dft.gov.uk/traffic-counts/index.php
Go to the interactive map and select location to be examined

Chapter 4

4.1 Page 57
Equality Act 2010. Public sector equality duty
https://www.legislation.gov.uk/ukpga/2010/15/section/149

4.2 Page 60
Walking & mental health
NHS Walking, dancing, cycling for health
http://www.nhs.uk/Livewell/getting-started-guides/Pages/getting-started-walking.aspx
Living Streets - Walk to school campaign
https://www.livingstreets.org.uk/what-we-do/walk-to-school

4.3 Page 61
Cycle Proofing Working Group
https://www.gov.uk/government/groups/cycle-proofing-working-group

4.4 Pages 61 & 68
Attachment of street lamps and signs to buildings
http://www.legislation.gov.uk/ukla/2013/5/section/4/enacted

4.5 page 66
Link between trees and mental health
http://nhsforest.org/evidence

4.6 Page 68
Code of Practice on Litter and Refuse, April 2006
Department for Environment Food and Rural Affairs
https://www.gov.uk/government/uploads/system/uploads/attachment_data/file/221087/pb11577b-cop-litter.pdf

Chapter 5

5.1 Page 73
Cost of road maintenance
The Economics of Road Maintenance. An RAC Foundation view
http://www.racfoundation.org/assets/rac_foundation/content/downloadables/economics_of_road_maintenance-an_racf_view-june_2013.pdf

5.2 Page 77
Finding your local highway authority
The website of a local highway authority with names of decision making councillors, highway policies, programmes, budgets, and current and future projects, can be found by entering a post-code into the national website: www.gov.uk/report-pothole

5.3 Page 78
Section 106 of Town and Country Planning Acts 1990
Department for Communities and Local Government
Section 106 Planning Obligations 2001 – 12. Published May 2014
https://www.gov.uk/government/uploads/system/uploads/attachment_data/file/314066/Section_106_Planning_Obligations_in_England_2011-12_-_Report_of_study.pdf
Section 278 of Highways Act 1980
http://www.legislation.gov.uk/ukpga/1980/66/section/278
Community Infrastructure Levy Act 2010
https://www.gov.uk/government/uploads/system/uploads/attachment_data/file/6313/1897278.pdf

5.4 Page 87
Interdisciplinary organisations
Academy of Urbanism: www.academyofurbanism.org.uk
Design Council: www.designcouncil.org.uk
Historic Towns Forum: www.historictownsforum.org
Place Alliance: placealliance.org.uk
Public Realm Information and Advice Network (PRIAN): www.PublicRealm.org & www.streetscapes.online
Town and Country Planning Association: www.tcpa.org.uk
Urban Design Group: www.udg.org.uk
Urban Design London: urbandesignlondon.com

Part ll

Part ll.1 Page 89
Creating better streets: Inclusive and accessible places
http://www.ciht.org.uk/en/document-summary/index.cfm/docid/BF28B40D-9855-46D6-B8C19E22B64AA066

Chapter 11

11.1 Page 133
Tourism spend in the Cotswolds
www.cotswoldaonb.org.uk Assessment of economic value of the Cotswold AONB, 2013. Cotswold sustainable tourism partnership

Updated references
At the time of writing, updated reference links are available on the PRIAN website: www.PublicRealm.org & www.streetscapes.online

**In addition to those mentioned
in the text, a sincere thank you to:**

Lynda Addison
Brenden Bell
Kate Carpenter
Alaine Coupre
Rob Cowan
Steven Cross
John Dales
Kate Davis
Fay Gibbons
Alison Gregory
Michael Heap
Peter Heath
Robert Huxford
Peter Jones

Phil Jones
Bob Leonard
Alex Luck
Jean Morgan
Mike Morris
Peter Piet
Stuart Reid
Anna Rose
Mahmood Sidiqui
Joanna Thomas
Robbie Thomas
Charles Wagner
Judith Walker
Sam Wright

Department for Transport

Image credits
Images are the copyright of the author except for:
Mrs Jaqueline Cullen: 0.01, 0.02, 0.03, 0.04
Surrey Hills AONB: 1.01
Look and Learn: 1.10
Christopher Thomas: 1.12
Munroe County Dep. of Tourism, Tennessee, USA: 1.15
Coventry City Council: 2.01, 7.04
Cleary Stevens Consulting: 2.18
Alex Luck: 3.16
Living Streets: 4.01
Capital & Counties Properties PLC: 4.02, 4.06, 4.14, 4.18, 5.01, 5.02, 5.08, 5.09, 5.10, 5.11
JC Decaux Ltd: 4.08
Northbank BID: 5.04, 5.05, 5.06,
Planit IE: 9.01, 9.04, 9.07, 9.09, 9.15
Hamilton-Baillie Ltd: 9.02, 9.06, 9.08
Robert Huxford: 9.13
Cotswold District Council: 11.03

**Professional and representative organisations
referred to in Fig. 5.03, page 75 & Box 5.2, page 86**

All-party Parliamentary Groups, UK Parliament: www.parliament.uk
Arboricultural Association: www.trees.org.uk
Association for Studies in the Conservation of Historic Buildings: www.aschb.org.uk
Association of Town & City Management: www.atcm.org
Business Improvement Districts: www.britishbids.info
British Cleaning Council: www.britishcleaningcouncil.org
British Ecological Society: www.britishecologicalsociety.org
British Psychological Society: www.bps.uk
Chartered Institution of Highways and Transportation: ciht.org.uk
Chartered Institution of Waste Management: www.ciwm.co.uk
Civic Voice: www.civicvoice.org.uk
Cycling: www.cycling.org & www.sustrans.org.uk
Commission for Global Road Safety: www.makeroadssafe.org
Confederation of Passenger Transport UK: www.cpt-uk.org
Highway Electrical Association: www.thehea.org.uk
Institute of Highways Engineers: www.theihe.org
Institute of Historic Building Conservation: www.ihbc.org.uk

Institute of Road Safety Officers: www.irso.org.uk & www.roadsafetygb.org.uk
Landscape Institute: www.landsacpeinstitute.org
Local Authority CEOs: www.solace.com
Local Authority Road Safety Officers' Association: www.rospa.com
Local Government Association: www.local.gov.uk
National Association of Disability Practitioners: www.nadp.org
National Association of Local Councils: www.nalc.gov.uk
National Market and Street Traders Association: www.nmtf.co.uk
Royal Institute of British Architects: www.architecture.com
Royal Institution of Chartered Surveyors: www.rics.org
Royal National Institute of Blind People: www.rnib.org.uk
Royal Town Planning Institute: www.rtpi.org.uk
Society of Local Council Clerks: www.slcc.co.uk
Transport Association: www.transportassociation.org.uk
Transport Focus: www.transportfocus.org.uk
Transport Planning Society: www.tps.org
Urban Art Association: www.urbanartsssociation.com
Walking: www.livingstreets.org.uk